INTERNATIONAL CENTRE FOR MECHANICAL SCIENCES

COURSES AND LECTURES No. 172

FORCED LINEAR VIBRATIONS

BY

PETER C. MÜLLER

AND

WERNER O. SCHIEHLEN

SPRINGER-VERLAG WIEN GMBH

Originally published by Springer-Verlag Wien-New York in 1977

ISBN 978-3-211-81487-1 ISBN 978-3-7091-4356-8 (eBook)
DOI 10.1007/978-3-7091-4356-8

PREFACE

This textbook contains, with some extensions, our lectures given at the Department of General Mechanics of the International Centre for Mechanical Sciences (CISM) in Udine/Italy during the month of October, 1973.

The book is divided into four major parts. The first part (Chapter 2, 3) is concerned with the mathematical representation of vibration systems and the corresponding general solution. The second part (Chapter 4) deals with the boundedness and stability of vibration systems. Thus, information on the general behavior of the system is obtained without any specified knowledge of the initial conditions and forcing functions. The third part (Chapter 5, 6) is devoted to deterministic excitation forces. In particular, the harmonic excitation leads to the phenomena of resonance, pseudoresonance and absorption. The fourth part (Chapter 7) considers stochastic excitation forces. The covariance analysis and the spectral density analysis of random vibrations are presented. Throughout the book examples are inserted for illustration.

In conclusion, we wish to express our gratitude to the International Centre for Mechanical Sciences (CISM) and to Professor Sobrero who invited us to deliver the lecture in Udine. We also acknowledge the support of Professor Magnus from the Institute B of Mechanics at the Technical University Munich.

Munich, October 1973

Peter C. Müller *Werner O. Schiehlen*

CHAPTER 1

Introduction

The subject of vibration deals with the oscillatory behavior of physical systems. The interaction of mass and elasticity allows vibration as well as the interaction of induction and capacity. Most vehicles, machines and circuits experience vibration and their design generally requires consideration of their oscillatory behavior.

Vibration systems can be characterized as linear or non-linear, as time-invariant or time-variant, as free or forced, as single-degree of freedom or multi-degree of freedom. For linear systems the principle of superposition holds, and the mathematical techniques available for their treatment are well-developed in matrix and control theory. In contrast, for the analysis of nonlinear systems the techniques are only partially developed and they are based mainly on approximation methods. For linear, time-invariant systems the concept of modal analysis is available featuring eigenvalues and eigenvectors. In contrary, for the analysis of linear, time-variant systems the fundamental matrix has to be found by numerical integration. Free vibrations take place when a system oscillates without external impressed forces. The system under free vibration will oscillate at its natural frequencies or eigenfrequencies. In contrast, forced vibrations take place under the excitation of external forces, in particular, impulse, periodic and stochastic forces. Single-degree of freedom systems are characterized by a scalar differential equation of second order. In contrary, multi-degree of freedom systems are usually described by vector and matrix differential equations. The number of degrees of freedom is equal to the minimum number of generalized coordinates necessary to describe the motion of the system. In addition to the notions presented above, Magnus (1969) uses the notions self-excited and parameter-excited. Self-excited vibrations may occur in nonlinear time-invariant, free systems while parameter-excited vibrations are typical for linear, periodic time-invariant free systems.

In this contribution, linear, time-invariant forced vibrations of mechanical systems with multi-degrees of freedom will be considered. Linear time-invariant systems are often obtained by the linearization of mechanical systems in the neighborhood of an equilibrium position. Forced systems result in addition to free systems in many vital phenomena such as resonance, pseudo-resonance, absorption and random vibrations. Multi-degree of freedom systems are usually necessary for an adequate representation of mechanical systems. Even if a continuous elastic body has an infinite number of degrees of freedom, in many cases, part of such bodies may be assumed to be rigid and the system may be dynamically equivalent to one with finite degrees of freedom.

A rigorous treatment is given to the boundedness and stability of the system's vibration, to resonances including pseudo-resonance and absorption, and to the random vibration analysis via the covariance and the spectral density technique. The computer-minded matrix theory is applied and approved numerical algorithms are mentioned to serve the special needs of multi-degree of freedom systems. But simple examples are also analytically treated to achieve a better understanding.

CHAPTER 2

Mathematical Representation of Mechanical Vibration Systems

The mathematical representation of a mechanical system requires firstly an adequate model. Secondly, one of the principles of dynamics has to be applied to the model and, then, the equations of motion are obtained. Finally, the equations of motion can be summarized to the state equation of the vibration system.

2.1 Modeling of Vibration Systems

For the modeling of vibration systems four approaches can be listed:

1. Multi-body approach,
2. Finite element approach,
3. Continuous system approach,
4. Hybrid approach.

For each engineering problem, the appropriate approach has to be elected. Some examples may illustrate the proceeding.

The vibrations of an automobile suspension can be properly modeled by a three-body system, Fig. 2.1, where the automobile body and the wheels and axles are considered as rigid bodies connected by springs and dashpots.

Further, the elasticity of the tires is represented by springs without damping.

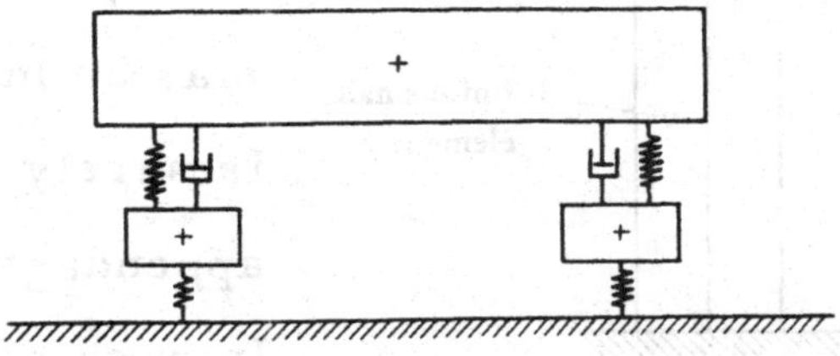

Fig.2.1. Three-body model of an automobile suspension

The vibrations of a spinning centrifuge with respect to its flexible suspension can be modeled by a rotating rigid body in the best manner, Fig. 2.2. The suspension is represented by spring and dashpot.

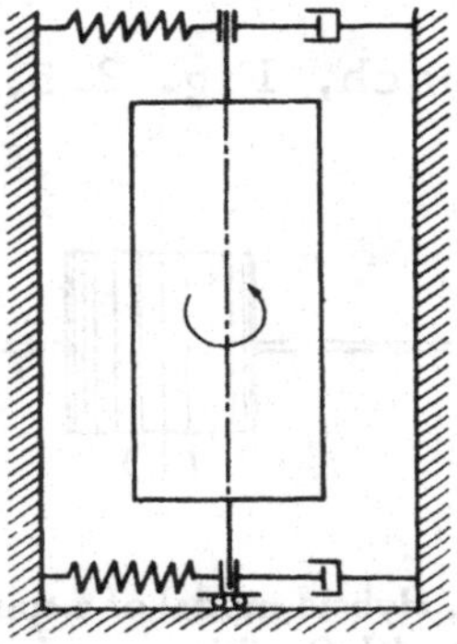

Fig.2.2. One-body model of a centrifuge

The bending vibrations of an automobile body have to be modeled by bar, rectangular and triangular elements, Fig. 2.3. Each element is considered as a flexible body where stiffness, damping and mass are concentrated in the nodes connecting the elements.

The torsional vibrations of a uniform bar are modeled best by a continuous system, Fig. 2.4. The infinite small elements are furnished with mass and elasticity.

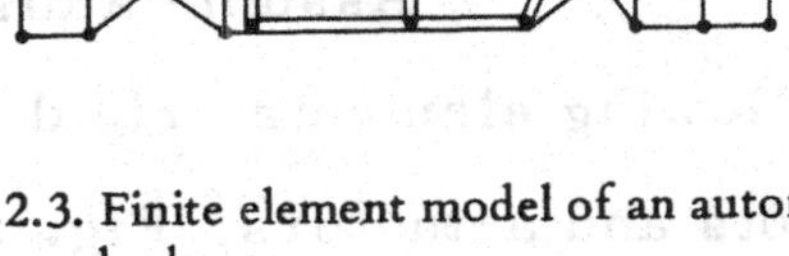

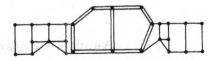

Fig.2.3. Finite element model of an automobile body

However, sometimes the three fundamental

approaches do not fit the engineering problem as well. As an example the spinning flexible satellite may be mentioned. Here, the core body is surely a rigid body while the flexible appendages represent continuous bars. In such cases, the continuous system may be replaced by a large number of elastically interconnected rigid bodies.

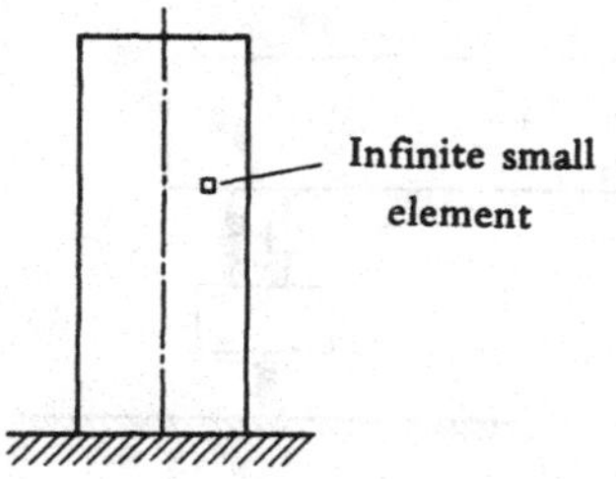

Fig.2.4. Continuous system model of a bar

Then, the multi-body approach can be used again. Or a hybrid approach, Fig. 2.5, is used combining the multi-body and the continuous system approach; see Likins (1971).

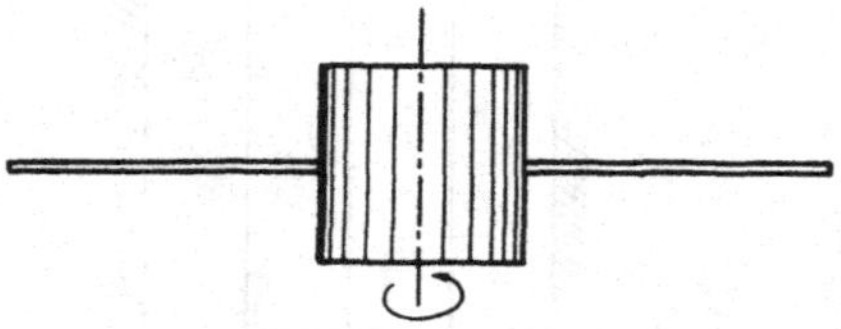

Fig.2.5. Hybrid model of a spinning satellite with flexible appendages

In the next sections the three fundamental approaches will be reviewed in short and the corresponding principles of dynamics will be applied.

2.2 Multi-body Approach

Assume a discrete, mechanical system with the following elements: rigid bodies with constraints, springs dashpots and actuators, Fig. 2.6. Then, either Euler's equation together with Newton's equation or Lagrange's equation may be applied. Both methods require the same kinematics.

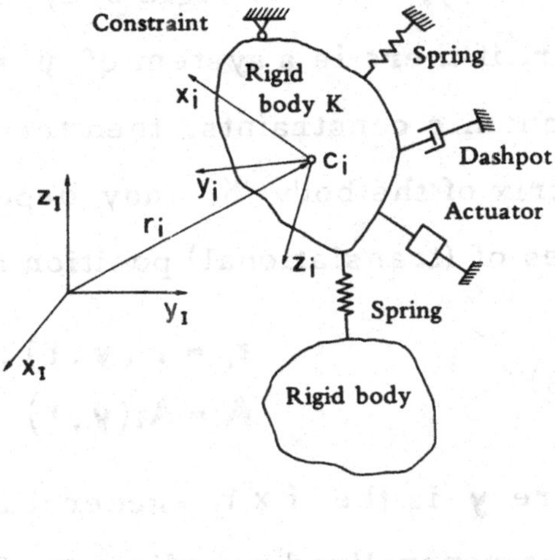

Fig.2.6. Discrete mechanical system with rigid bodies

<u>Kinematics</u>

The position of the rigid body K_i is uniquely characterized in space by a body-fixed, orthogonal frame. With respect to the inertial frame x_I, y_I, z_I, the body-fixed frame x_i, y_i, z_i, with origin at the center of mass C_i can be described by the 3×1-position vector r_i and the 3×3-rotation matrix A_i. If there is only one free rigid body, then the position vector may be given by three Cartesian coordinates

$$r_i = \begin{bmatrix} r_x & r_y & r_z \end{bmatrix}^T , \quad i = 1 , \tag{2.1}$$

and the rotation matrix may be given by three Euler angles, representing three generalized coordinates,

$$A_i = \begin{bmatrix} \cos\theta \cos\psi & -\cos\theta \sin\psi & \sin\theta \\ \cos\phi \sin\psi + \sin\phi \sin\theta \cos\psi & \cos\phi \cos\psi - \sin\phi \sin\theta \sin\psi & -\sin\phi \cos\theta \\ \sin\phi \sin\psi - \cos\phi \sin\theta \cos\psi & \sin\phi \cos\psi + \cos\phi \sin\theta \sin\psi & \cos\phi \cos\theta \end{bmatrix} , i = 1. \tag{2.2}$$

Obviously, the free rigid body has six degrees of freedom. However, if there is a system of p rigid bodies, possibly with some holonomic constraints, then the position vector and the rotation matrix of the body K_i may depend on all generalized coordinates of (translational) position as well as of rotation

(2.3) $$\left.\begin{aligned} \mathbf{r}_i &= \mathbf{r}_i(\mathbf{y}, t), \\ \mathbf{A}_i &= \mathbf{A}_i(\mathbf{y}, t), \end{aligned}\right\} i = 1(1)p$$

where $\mathbf{y}$ is the $f \times 1$ -generalized position vector composed of the generalized coordinates. For the system's number of degrees of freedom it yields

(2.4) $$f \leq 6p .$$

The 3×1 -velocity vector $\mathbf{v}_i$ and the 3×1 -angular velocity vector $\boldsymbol{\omega}_i$ of the body K_i with respect to the inertial frame are obtained by differentiation of (2.3)

(2.5) $$\left.\begin{aligned} \mathbf{v}_i &= \mathbf{J}_{Ti}\,\dot{\mathbf{y}} + \bar{\mathbf{v}}_i, & \bar{\mathbf{v}}_i &= \partial \mathbf{r}_i / \partial t, \\ \boldsymbol{\omega}_i &= \mathbf{J}_{Ri}\,\dot{\mathbf{y}} + \bar{\boldsymbol{\omega}}_i, & \bar{\boldsymbol{\omega}}_i &= \partial \mathbf{a}_i / \partial t, \end{aligned}\right\} i = 1(1)p$$

where

(2.6) $$\mathbf{J}_{Ti} = \frac{\partial \mathbf{r}_i}{\partial \mathbf{y}} = \begin{bmatrix} \frac{\partial r_{xi}}{\partial y_1} & \frac{\partial r_{xi}}{\partial y_2} & \cdots & \frac{\partial r_{xi}}{\partial y_f} \\ \frac{\partial r_{yi}}{\partial y_1} & \frac{\partial r_{yi}}{\partial y_2} & \cdots & \frac{\partial r_{yi}}{\partial y_f} \\ \frac{\partial r_{zi}}{\partial y_1} & \frac{\partial r_{zi}}{\partial y_2} & \cdots & \frac{\partial r_{zi}}{\partial y_f} \end{bmatrix}, \quad i = 1(1)p$$

is the $3 \times f$ -Jacobian matrix of translation and

$$\mathfrak{J}_{Ri} = \frac{\partial a_i}{\partial y} = \begin{bmatrix} \frac{\partial a_{xi}}{\partial y_1} & \frac{\partial a_{xi}}{\partial y_2} & \cdots & \frac{\partial a_{xi}}{\partial y_f} \\ \frac{\partial a_{yi}}{\partial y_1} & \frac{\partial a_{yi}}{\partial y_2} & \cdots & \frac{\partial a_{yi}}{\partial y_f} \\ \frac{\partial a_{zi}}{\partial y_1} & \frac{\partial a_{zi}}{\partial y_2} & \cdots & \frac{\partial a_{zi}}{\partial y_f} \end{bmatrix}, \quad i = 1(1)p \qquad (2.7)$$

is the $3 \times f$ -Jacobian matrix of rotation. The angular velocity $\bar{\omega}_i$ and the rotational Jacobian matrix $\mathfrak{J}_{Ri}$ are obtained from the corresponding skew-symmetric rotation tensors

$$\frac{\partial \tilde{a}_i}{\partial t} = \frac{\partial \mathbf{A}_i}{\partial t} \cdot \mathbf{A}_i^T, \quad \frac{\partial \tilde{a}_i}{\partial y_j} = \frac{\partial \mathbf{A}_i}{\partial y_j} \mathbf{A}_i^T, \quad \begin{matrix} i = 1(1)p \\ j = 1(1)f \end{matrix} \qquad (2.8)$$

where

$$\tilde{\mathbf{a}}_i = \begin{bmatrix} 0 & -a_{zi} & a_{yi} \\ a_{zi} & 0 & -a_{xi} \\ -a_{yi} & a_{xi} & 0 \end{bmatrix} \quad \text{for} \quad \mathbf{a}_i = \begin{bmatrix} a_{xi} \\ a_{yi} \\ a_{zi} \end{bmatrix} \qquad (2.9)$$

and $\mathbf{a}_i$ is a 3×1 -vector. Thus, $\sim$ characterizes the matrix notation of the vector cross product.

Newton's and Euler's Equation

Newton's equation reads for each rigid body K_i with respect to the center of mass C_i as

$$m_i \dot{\mathbf{v}}_i = \mathbf{f}_i, \quad i = 1(1)p \qquad (2.10)$$

where m_i is the scalar mass and $\mathbf{f}_i$ is the 3×1 -force vector including all forces acting on body K_i. Euler's equation reads for each body K_i with respect to C_i as

(2.11) $$\mathbf{I}_i \dot{\boldsymbol{\omega}}_i + \tilde{\boldsymbol{\omega}}_i \mathbf{I}_i \boldsymbol{\omega}_i = \mathbf{l}_i , \quad i = 1(1)p$$

where $\mathbf{I}_i$ is the 3×3 inertia tensor of body K_i and $\mathbf{l}_i$ is the 3×1 -torque vector including all torques acting on body K_i. The force $\mathbf{f}_i$ and the torque $\mathbf{l}_i$ depend in forced vibration systems on the generalized coordinates (spring forces), on the generalized velocities (dashpot forces), on the time (actuator forces) and on the constraints

(2.12) $$\left.\begin{aligned} \mathbf{f}_i &= \mathbf{f}_{Bi}(\mathbf{y}, \dot{\mathbf{y}}, t) + \mathbf{f}_{Ci} , \\ \mathbf{l}_i &= \mathbf{l}_{Bi}(\mathbf{y}, \dot{\mathbf{y}}, t) + \mathbf{l}_{Ci} , \end{aligned}\right\} \quad i = 1(1)p$$

where $\mathbf{f}_{Ci}$, $\mathbf{l}_{Ci}$ are due to the constraints.
Introducing (2.5) and (2.12) in (2.10),(2.11) it remains

(2.13) $$\left\{\begin{aligned} & m_i \mathbf{J}_{Ti} \ddot{\mathbf{y}} + m_i \dot{\mathbf{J}}_{Ti} \dot{\mathbf{y}} + m_i \dot{\bar{\mathbf{v}}}_i = \mathbf{f}_{Bi}(\mathbf{y}, \dot{\mathbf{y}}, t) + \mathbf{f}_{Ci} , \\ & \mathbf{I}_i \mathbf{J}_{Ri} \ddot{\mathbf{y}} + \mathbf{I}_i \dot{\mathbf{J}}_{Ri} \dot{\mathbf{y}} + \mathbf{I}_i \dot{\bar{\boldsymbol{\omega}}}_i + \left(\widetilde{\mathbf{J}_{Ri} \dot{\mathbf{y}} + \bar{\boldsymbol{\omega}}_i}\right) \mathbf{I}_i \left(\mathbf{J}_{Ri} \dot{\mathbf{y}} + \bar{\boldsymbol{\omega}}_i\right) = \\ & = \mathbf{l}_{Bi}(\mathbf{y}, \dot{\mathbf{y}}, t) + \mathbf{l}_{Ci} , \\ & i = 1(1)p \end{aligned}\right.$$

The $6p$ scalar equations (2.13) can be summarized in matrix notation

(2.14) $$\bar{\mathbf{M}}(\mathbf{y}, t) \ddot{\mathbf{y}} + \bar{\mathbf{g}}(\dot{\mathbf{y}}, \mathbf{y}, t) + \bar{\mathbf{f}}_C = \mathbf{0}$$

where $\bar{\mathbf{M}}$ is a $6p \times f$ -mass matrix, $\bar{\mathbf{g}}$ is a $6p \times 1$ -vector function including $\bar{\mathbf{v}}_i$ and $\bar{\boldsymbol{\omega}}_i$ and $\mathbf{f}_C$ is the $6p \times 1$ -vector of the constraint forces and torques. Thus, one gets $6p$ equations for the f generalized coordinates and $6p-f$ linear independent constraint forces. Usually, however, the constraint forces are not required and for system order reduction they have, then, to be eliminated. This can be done by the principle of virtual work regarding (2.6), (2.7):

$$\sum_{i=1}^{p}\left(\mathbf{f}_{Ci}^T\,\delta\mathbf{r}_i + \mathbf{l}_{Ci}^T\,\delta\mathbf{a}_i\right) = \delta\mathbf{y}^T\sum_{i=1}^{p}\left(\mathbf{J}_{Ti}^T\,\mathbf{f}_{Ci} + \mathbf{J}_{Ri}^T\,\mathbf{l}_{Ci}\right) = 0 \qquad (2.15)$$

or

$$\bar{\mathbf{J}}^T\,\bar{\mathbf{f}}_C = 0 \qquad (2.16)$$

where $\bar{\mathbf{J}}^T = \left[\bar{\mathbf{J}}_{T_1}^T\ \bar{\mathbf{J}}_{T_2}^T \cdots\cdots \bar{\mathbf{J}}_{Rp-1}^T\ \bar{\mathbf{J}}_{Rp}^T\right]$ is the global $f \times 6p$-Jacobian matrix. Then, premultiplying (2.14) by $\bar{\mathbf{J}}^T$, it remains

$$\mathbf{M}(\mathbf{y},t)\,\ddot{\mathbf{y}} + \mathbf{g}(\dot{\mathbf{y}},\mathbf{y},t) = 0 \qquad (2.17)$$

where $\mathbf{M}$ is the $f \times f$ -symmetric mass matrix and $\mathbf{g}$ is a $f \times 1$ -vector function.

Lagrange's equation

Lagrange's equation reads for a system of p rigid bodies as

$$\frac{d}{dt}\,\frac{\partial T}{\partial \dot{\mathbf{y}}} + \frac{\partial T}{\partial \mathbf{y}} = \mathbf{q}\,, \qquad (2.18)$$

where

$$T = \frac{1}{2}\sum_{i=1}^{p}\left(\mathbf{v}_i^T\,m_i\,\mathbf{v}_i + \boldsymbol{\omega}_i^T\,\mathbf{I}_i\,\boldsymbol{\omega}_i\right) \qquad (2.19)$$

is the scalar kinetic energy and

$$\mathbf{q} = \sum_{i=1}^{p} \left(\mathfrak{J}_{Ti}^T \mathbf{f}_i + \mathfrak{J}_{Ri}^T \mathbf{l}_i \right) \tag{2.20}$$

is the generalized $f \times 1$ -force vector.

As simple as Lagrange's equation is looking as difficult may be the evaluation. This is obvious if (2.5) is introduced in (2.19)

$$\begin{aligned} T = \frac{1}{2} \sum_{i=1}^{p} \Big(& \dot{\mathbf{y}}^T \mathfrak{J}_{Ti}^T m_i \mathfrak{J}_{Ti} \dot{\mathbf{y}} + 2 \dot{\mathbf{y}} \mathfrak{J}_{Ti} m_i \bar{\mathbf{v}}_i + \bar{\mathbf{v}}_i^T m_i \bar{\mathbf{v}}_i + \\ & + \dot{\mathbf{y}}^T \mathfrak{J}_{Ri}^T \mathbf{I}_i \mathfrak{J}_{Ri} \dot{\mathbf{y}} + 2 \dot{\mathbf{y}} \mathfrak{J}_{Ri} \mathbf{I}_i \bar{\boldsymbol{\omega}}_i + \bar{\boldsymbol{\omega}}_i^T \mathbf{I}_i \bar{\boldsymbol{\omega}}_i \Big) . \end{aligned} \tag{2.21}$$

However, after proceeding through the differentiation of the kinetic energy (2.21), it finally follows from Lagrange's equation (2.18) exactly the same equation of motion (2.17) as obtained via Newton's and Euler's equation. In recent days of digital and electronic computers, therefore, Newton's and Euler's equation seem to be more convenient since only matrix operations are required. Further, Euler's and Newton's equation can be easily extended to moving reference frames (relative motion) as shown by Schiehlen (1972) .

Linearization

Assume equilibrium position $\mathbf{y} = 0$ and small oscillations in the neighborhood of the equilibrium position. Then, the second and higher order terms in the generalized coordinates can be neglected and it remains from (2.17)

$$\mathbf{M}\ddot{\mathbf{y}} + (\mathbf{D}+\mathbf{G})\dot{\mathbf{y}} + (\mathbf{K}+\mathbf{N})\mathbf{y} = \mathbf{h}(t). \qquad (2.22)$$

Here, $\mathbf{M}$, $\mathbf{D}$ and $\mathbf{K}$ are symmetric $f \times f$ -matrices; $\mathbf{G}$, $\mathbf{N}$ are skew-symmetric $f \times f$ -matrices and $\mathbf{h}(t)$ is a $f \times 1$ -forcing vector.

2.3 Finite Element Approach

Assume a discrete, mechanical system with finite, flexible elements such as bars, triangles, cubes, etc., Fig. 2.7. Firstly, each element is characterized by n nodes in an arbitrary frame. Then, the stiffness, the internal viscous damping and the inertia of each element are concentrated at the nodes. Applying the principles of mechanics equivalent generalized forces are obtained for each element separately

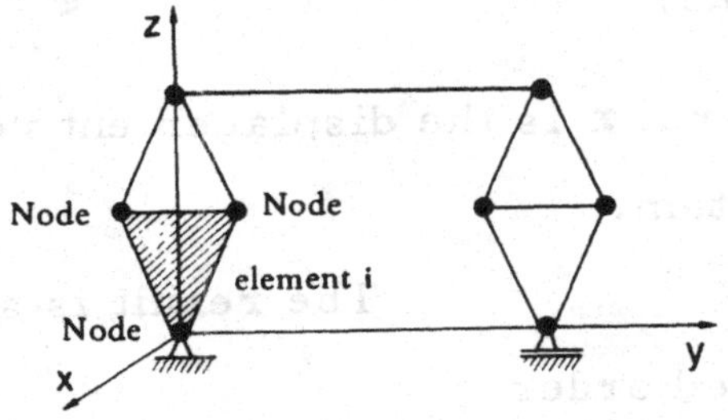

Fig.2.7. Discrete mechanical system with flexible elements

$$\left.\begin{aligned} \mathbf{f}_K &= \mathbf{k}\,\boldsymbol{\varrho}, \\ \mathbf{f}_D &= \mathbf{d}\,\dot{\boldsymbol{\varrho}}, \\ \mathbf{f}_M &= \mathbf{m}\ddot{\boldsymbol{\varrho}} \end{aligned}\right\} \qquad (2.23)$$

where $\mathbf{f}$ are $6n \times 1$ -force vectors, $\mathbf{k}, \mathbf{d}, \mathbf{m}$ are $6n \times 6n$ -matrices characterizing stiffness, damping and inertia, and ϱ is the $6n \times 1$ -displacement vector of the considered element. Further, there may be given additional (external) forces $\mathbf{f}(t)$ acting on some nodes. Secondly, the forces at all nodes of all elements are summarized in a global matrix equation

$$\hat{\mathbf{M}} \ddot{\hat{\varrho}} + \hat{\mathbf{D}} \dot{\hat{\varrho}} + \hat{\mathbf{K}} \hat{\varrho} = \hat{\mathbf{f}} \tag{2.24}$$

where $\hat{\mathbf{M}}, \hat{\mathbf{D}}, \hat{\mathbf{K}}$ are diagonal-hyper-matrices and $\hat{\varrho}, \hat{\mathbf{f}}$ are hyper-vectors of corresponding dimension. Thirdly, the kinematic constraints between the elements are regarded by an incidence matrix $\mathbf{A}$:

$$\hat{\varrho} = \mathbf{A} \mathbf{z} \tag{2.25}$$

where $\mathbf{z}$ is the displacement vector of the considered (free) system.

The result is a matrix equation of strongly reduced order

$$\bar{\bar{\mathbf{M}}} \ddot{\mathbf{z}} + \bar{\bar{\mathbf{D}}} \dot{\mathbf{z}} + \bar{\bar{\mathbf{K}}} \mathbf{z} = \bar{\bar{\mathbf{f}}} . \tag{2.26}$$

Then, the boundary conditions have to be considered, i.e. the displacement of some essential nodes may be constrained while the corresponding forces are unknown. After the elimination of these reaction forces, finally the equation of motion is obtained:

$$\mathbf{M}\ddot{\mathbf{y}} + \mathbf{D}\dot{\mathbf{y}} + \mathbf{K}\mathbf{y} = \mathbf{h}(t) \tag{2.27}$$

where $\mathbf{y}$ is the $f \times 1$ -vector of the generalized node coordinates; $\mathbf{M}, \mathbf{D}, \mathbf{K}$ are symmetric $f \times f$ -matrices and $\mathbf{h}(t)$ is the $f \times 1$ -forcing function. For more details of the finite element method, usually applied to vibrations of structures, see Zienkiewicz (1971).

2.4 Continuous System Approach

Assume a continuous mechanical system, Fig. 2.8, without any obvious element. Then, for an infinite small element of the continuous system, a partial differential equation can be obtained. in general, the equation reads for a two-dimensional space as

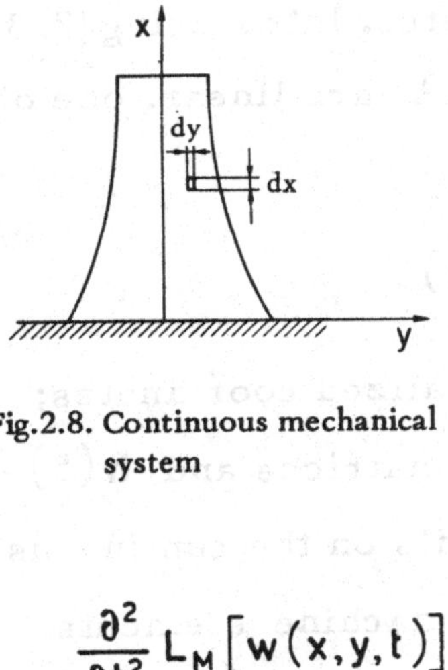

Fig.2.8. Continuous mechanical system

$$\frac{\partial^2}{\partial t^2} L_M\left[w(x,y,t)\right] + \frac{\partial}{\partial t} L_D\left[w(x,y,t)\right] + L_K\left[w(x,y,t)\right] =$$

$$= F(x,y,t) \tag{2.28}$$

where

$$L = A_1 + A_2 \frac{\partial}{\partial x} + A_3 \frac{\partial}{\partial y} + A_4 \frac{\partial^2}{\partial x^2} + \ldots \tag{2.29}$$

is a linear differential operator, $w(x,y)$ is the displacement at the position (x,y) and $F(x,y,t)$ is the force acting at (x,y). In addition, the boundary conditions have to be satisfied, i.e.,

$$B\left[w(x,y,t)\right] = 0 \tag{2.30}$$

where B is another linear differential operator. Then, the solution is approximated by a series

$$w(x,y,t) = \sum_{r=1}^{f} w_r(x,y)\, y_r(t) \tag{2.31}$$

where $w_r(x,y)$ are the eigenfunctions of the free, undamped system and $y_r(t)$ are generalized coordinates. Introducing (2.31) in (2.28) and recalling that the operators L are linear, one obtains the matrix differential equation

$$\mathbf{M}\ddot{\mathbf{y}} + \mathbf{D}\dot{\mathbf{y}} + \mathbf{K}\mathbf{y} = \mathbf{h}(t) \tag{2.32}$$

where $\mathbf{y}$ is the $f \times 1$-vector of the generalized coordinates; $\mathbf{M}, \mathbf{D}, \mathbf{K}$ are constant, symmetric $f \times f$-matrices and $\mathbf{h}(t)$ is a $f \times 1$-forcing vector. For more details on the continuous system approach, often applied to simple machine elements, see Meirovitch (1967).

2.5 State Equation of Vibration Systems

The equation of motion of a forced linear vibration system obtained in the previous three sections has always the typical form featuring the second order equation

$$\left.\begin{aligned} \mathbf{M}\ddot{\mathbf{y}}(t) + (\mathbf{D}+\mathbf{G})\dot{\mathbf{y}}(t) + (\mathbf{K}+\mathbf{N})\mathbf{y}(t) = \mathbf{h}(t), \\ \dot{\mathbf{y}}(t_0) = \dot{\mathbf{y}}_0, \quad \mathbf{y}(t_0) = \mathbf{y}_0 \end{aligned}\right\} \qquad (2.33)$$

where $\mathbf{y}(t)$ is the $f \times 1$ -vector of the generalized coordinates, $\mathbf{h}(t)$ is the $f \times 1$ -vector of the forcing function and $\mathbf{M}$, $\mathbf{D}$, $\mathbf{G}$, $\mathbf{K}$, $\mathbf{N}$ are $f \times f$ -matrices. Further, the initial conditions $\mathbf{y}(t_0)$ and $\dot{\mathbf{y}}(t_0)$ are written for completeness. The matrices in (2.33) can also be interpreted physically: $\mathbf{M} = \mathbf{M}^T$ is the symmetric mass or inertia matrix, $\mathbf{D} = \mathbf{D}^T$ is the symmetric damping matrix usually due to viscous damping, $\mathbf{G} = -\mathbf{G}^T$ is the skew-symmetric gyro-matrix usually due to gyroscopic phenomena in rotating vibration systems, $\mathbf{K} = \mathbf{K}^T$ is the symmetric matrix of conservative forces usually due to springs and $\mathbf{N} = -\mathbf{N}^T$ is the skew-symmetric matrix of non-conservative forces sometimes due to damping in rotating systems. However, if there are active servomechanisms within the system, this interpretation may be not true. For constant matrices $\mathbf{M}, \mathbf{D}, \mathbf{G}, \mathbf{K}, \mathbf{N}$, the vibration system is called time-invariant. This will be assumed in the following throughout.

In addition to the second order equation, the first order state equation is essential

$$\dot{\mathbf{x}}(t) = \mathbf{A}\,\mathbf{x}(t) + \mathbf{f}(t)\ ,\quad \mathbf{x}(t_0) = \mathbf{x}_0 \tag{2.34}$$

where

$$\mathbf{x}(t) = \begin{bmatrix} \mathbf{y}(t) \\ \dot{\mathbf{y}}(t) \end{bmatrix} \tag{2.35}$$

is the $n \times 1$ -state vector,

$$\mathbf{f}(t) = \begin{bmatrix} \mathbf{0} \\ \mathbf{M}^{-1}\mathbf{h}(t) \end{bmatrix} \tag{2.36}$$

is the $n \times 1$ -forcing vector and

$$\mathbf{A} = \left[\begin{array}{c|c} \mathbf{O} & \mathbf{E} \\ \hline -\mathbf{M}^{-1}(\mathbf{K}+\mathbf{N}) & -\mathbf{M}^{-1}(\mathbf{D}+\mathbf{G}) \end{array}\right] \tag{2.37}$$

is the $n \times n$ -system matrix. Here, $\mathbf{O}$ is the $f \times f$ -zero matrix and $\mathbf{E}$ is the $f \times f$ -unit matrix. The dimension of the first order system is $n = 2f$. The first order state equation often allows the direct application of computer-minded algorithms for vibration analysis.

2.6 Examples

The state equations will be specified for an automobile suspension and a centrifuge using the multi-body approach. These examples will also be employed in the next chapters for illustration.

Example 2.1: Automobile wheel suspension

The automobile suspension is modeled by a two-body system, Fig. 2.9. Only vertical vibrations will be considered. Then, the small oscillations y_1, y_2 in the neighborhood of the equilibrium position are immediately generalized coordinates and Newton's equation (2.10) results in the following second order state equations

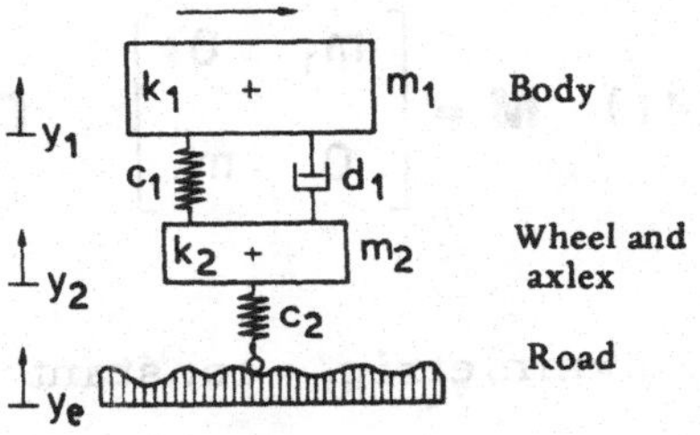

Fig.2.9. Automobile wheel suspension model

$$\left.\begin{aligned} m_1 \ddot{y}_1 &= -c_1(y_1 - y_2) - d_1(\dot{y}_1 - \dot{y}_2), \\ m_2 \ddot{y}_2 &= c_1(y_1 - y_2) + d_1(\dot{y}_1 - \dot{y}_2) - c_2(y_2 - y_e) \end{aligned}\right\} \quad (2.38)$$

The parameters m_i, c_i, d_i, $i = 1,2$ follow from Fig. 2.9 and the road is generating the forcing coordinate $y_e(t)$. Introducing the vector

(2.39) $$\mathbf{y} = \left[y_1 \;\; y_2 \right]^T$$

and the forcing vector

(2.40) $$\mathbf{h} = \left[0 \;\; c_2 y_e \right]^T$$

the matrix equation (2.33) is obtained

$$\mathbf{M}\ddot{\mathbf{y}}(t) + \mathbf{D}\dot{\mathbf{y}}(t) + \mathbf{K}\mathbf{y}(t) = \mathbf{h}(t)$$

where

(2.41) $$\mathbf{M} = \begin{bmatrix} m_1 & 0 \\ 0 & m_2 \end{bmatrix}, \quad \mathbf{D} = \begin{bmatrix} d_1 & -d_1 \\ -d_1 & d_1 \end{bmatrix}, \quad \mathbf{K} = \begin{bmatrix} c_1 & -c_1 \\ -c_1 & c_1 + c_2 \end{bmatrix}$$

are symmetric, constant 2 x 2 -matrices. Further, the first order state equation (2.34) reads as

$$\dot{\mathbf{x}}(t) = \mathbf{A}\mathbf{x}(t) + \mathbf{f}(t)$$

where

(2.42) $$\mathbf{x}(t) = \left[y_1 \;\; y_2 \;\; \dot{y}_1 \;\; \dot{y}_2 \right]^T,$$

(2.43) $$\mathbf{f}(t) = \left[0 \;\; 0 \;\; 0 \;\; \frac{c_2}{m_2} y_e \right],$$

(2.44) $$\mathbf{A} = \left[\begin{array}{cc|cc} 0 & 0 & 1 & 0 \\ 0 & 0 & 0 & 1 \\ \hline -\frac{c_1}{m_1} & \frac{c_1}{m_1} & -\frac{d_1}{m_1} & \frac{d_1}{m_1} \\ \frac{c_1}{m_2} & -\frac{c_1 + c_2}{m_2} & \frac{d_1}{m_2} & -\frac{d_1}{m_2} \end{array} \right].$$

Consider that the submatrices of $\mathbf{A}$ in (2.44) are non-symmetric.

Example 2.2: Centrifuge

The centrifuge is modeled by one rotating body, Fig. 2.10. The center of mass C is assumed to be fixed within the bearing. Then, the small Euler angles $\Phi, \Theta, \tilde{\psi}$ are generalized coordinates while the large Euler angle $\psi = \Omega t$ represents the spin motion where Ω is the constant spin rate. The rotation matrix (2.2) reads as

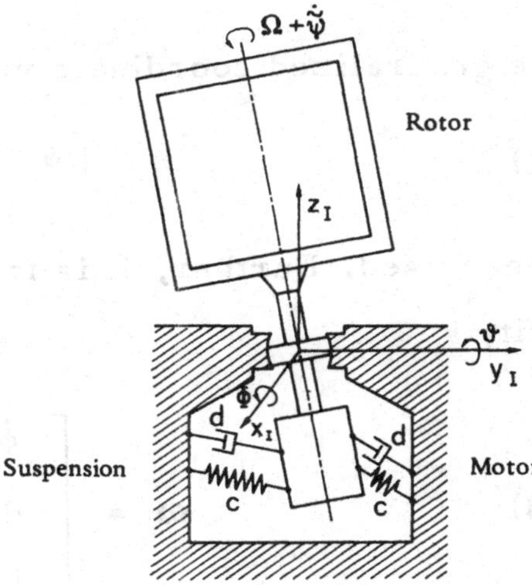

Fig.2.10. Centrifuge

$$\mathbf{A}_1 = \begin{bmatrix} \left(1-\frac{1}{2}\Theta^2\right)\cos\Omega t & -\left(1-\frac{1}{2}\Phi^2\right)\sin\Omega t & \Theta \\ \left(1-\frac{1}{2}\Phi^2\right)\sin\Omega t + \Phi\Theta\cos\Omega t & \left(1-\frac{1}{2}\Phi^2\right)\cos\Omega t - \Phi\Theta\sin\Omega t & -\Phi \\ \Phi\sin\Omega t - \Theta\cos\Omega t & \Phi\cos\Omega t + \Theta\sin\Omega t & 1-\frac{1}{2}\Phi^2-\frac{1}{2}\Theta^2 \end{bmatrix},$$

$$\Phi, \Theta \ll 1 \qquad (2.45)$$

The Jacobian matrix (2.7) follows as

$$\mathfrak{J}_R = \begin{bmatrix} 1 & 0 & \theta \\ 0 & 1 & -\phi \\ 0 & \phi & 1 \end{bmatrix} \tag{2.46}$$

where generalized coordinate vector

$$\eta = \begin{bmatrix} \phi & \theta & \tilde{\psi} \end{bmatrix}^T \tag{2.47}$$

has been used. Further, it is from (2.45) by (2.5) the angular velocity

$$\omega = \begin{bmatrix} \dot{\phi} + \Omega\theta \\ \dot{\theta} - \Omega\phi \\ \Omega + \dot{\tilde{\psi}} \end{bmatrix} . \tag{2.48}$$

Now, Euler's equation has to be applied. The inertia tensor of the symmetric centrifuge is in the body-fixed frame 1 assumed as

$${}_1\mathbf{I} = \begin{bmatrix} I_x & 0 & 0 \\ 0 & I_x & I_{yz} \\ 0 & I_{yz} & I_z \end{bmatrix}, \quad I_{yz} \ll I_x, I_z . \tag{2.49}$$

In the inertial frame I it yields

$$\mathbf{I} = \mathbf{A}_1 \cdot_1 \mathbf{I} \cdot \mathbf{A}_1^T = \begin{bmatrix} I_x & 0 & (I_z - I_x)\theta - I_{yz}\sin\Omega t \\ 0 & I_x & -(I_z - I_x)\phi + I_{yz}\cos\Omega t \\ (I_z - I_x)\theta - I_{yz}\sin\Omega t & -(I_z - I_x)\phi + I_{yz}\cos\Omega t & I_z \end{bmatrix} \tag{2.50}$$

where the transformation law for tensors in Cartesian frames is used. Further, the external torques by spring and dashpot of the suspension are given by

$$\mathbf{l} = -\begin{bmatrix} c\phi \\ c\theta \\ 0 \end{bmatrix} - \begin{bmatrix} d\dot{\phi} \\ d\dot{\theta} \\ 0 \end{bmatrix} . \tag{2.51}$$

Introducing (2.48), (2.50) and (2.51) in (2.11) it remains

$$\left.\begin{aligned} I_x\ddot{\phi} + d\dot{\phi} + c\phi + I_z\Omega\dot{\theta} &= I_{yz}\Omega^2\cos\Omega t, \\ I_x\ddot{\theta} + d\dot{\theta} + c\theta - I_z\Omega\dot{\phi} &= I_{yz}\Omega^2\sin\Omega t, \\ I_z \cdot 0 &= 0. \end{aligned}\right\} \tag{2.52}$$

Premultiplying (2.52) by $\mathfrak{J}_R^T$, the first two equations of (2.52) are obtained as essential equations. In matrix form, it follows with $\mathbf{y} = [\phi \ \ \theta]^T$

$$\ddot{\mathbf{y}}(t) + (\delta\mathbf{E} + \omega\mathbf{S})\dot{\mathbf{y}}(t) + \mathbf{k}\mathbf{y}(t) = \mathbf{h}(t) \tag{2.53}$$

where the abbreviations

$$k = \frac{c}{I_x}\ , \quad \delta = \frac{d}{I_x}\ , \quad \omega = \frac{I_z \Omega}{I_x}\ , \tag{2.54}$$

the skew-symmetric matrix

$$\mathbf{S} = \begin{bmatrix} 0 & 1 \\ -1 & 0 \end{bmatrix} \tag{2.55}$$

and the periodic forcing vector

$$\mathbf{h}(t) = \begin{bmatrix} \frac{I_{yz}}{I_x} \Omega^2 \cos \Omega t \\ \frac{I_{yz}}{I_z} \Omega^2 \sin \Omega t \end{bmatrix} \tag{2.56}$$

are used.

The first order state equation is trivial since $\mathbf{M} = \mathbf{E}$ and it will not be listed here.

CHAPTER 3

General Solution of Linear Vibration Systems

As shown in the previous chapter each finite dimensional linear (or linearized) dynamical mechanical system can be represented in state space notation by a set of n first order differential equations

$$\dot{\mathbf{x}}(t) = \mathbf{A}(t)\mathbf{x}(t) + \mathbf{f}(t) , \quad \mathbf{x}(t_0) = \mathbf{x}_0 , \tag{3.1}$$

where $\mathbf{x}$ is the $n \times 1$ state vector, $\mathbf{A}$ the $n \times n$ system matrix, and $\mathbf{f}$ the $n \times 1$ forcing vector. In this chapter the general solution of (3.1) and its properties are briefly reviewed in the case of a time-invariant system matrix $\mathbf{A}(t) = \mathbf{A}$ = constant.

3.1 Solution by the State Transition Matrix

The linear time invariant homogeneous system

$$\dot{\mathbf{x}}(t) = \mathbf{A}\mathbf{x}(t) \tag{3.2}$$

has a fundamental set of n linearly independent solutions $\boldsymbol{\varphi}_1(t), \ldots, \boldsymbol{\varphi}_n(t)$ satisfying the differential equation (3.2),

$$\dot{\boldsymbol{\varphi}}_i(t) = \mathbf{A}\boldsymbol{\varphi}_i(t) , \quad i = 1(1)n , \tag{3.3}$$

and the initial conditions

$$\varphi_i(0) = e_i \ , \quad i = 1(1)n \ , \tag{3.4}$$

where e_i is the i-th unit vector of the n-dimensional vector space. Arranging these solutions $\varphi_i(t)$ as the n columns of a matrix the state transition matrix (fundamental matrix) for (3.2) is obtained

$$\Phi(t) = \left[\varphi_1(t)\,\varphi_2(t)\ \ldots\ \varphi_n(t)\right] \tag{3.5}$$

Obviously, $\Phi(t)$ satisfies the matrix differential equation

$$\dot{\Phi}(t) = A\,\Phi(t) \ , \quad \Phi(0) = E \ . \tag{3.6}$$

The unique solution of (3.6) is formally given by a converging infinite power series defining the exponential function of the matrix At:

$$\Phi(t) = e^{At} = \sum_{i=0}^{\infty} \frac{(At)^i}{i!} \ . \tag{3.7}$$

Then, following properties of $\Phi(t)$ can be proved

$$\Phi(t_1+t_2) = \Phi(t_1)\,\Phi(t_2) \ , \tag{3.8}$$

$$\Phi^{-1}(t) = \Phi(-t) \ , \tag{3.9}$$

$$\det \Phi(t) = e^{t \cdot \mathrm{tr} A} \tag{3.10}$$

where det and tr means determinant and trace of a matrix.

Using the fundamental matrix (3.5) the general solution of (3.2) can be written as

$$x(t) = \sum_{i=1}^{n} \varphi_i(t)\, x_{i0} = \Phi(t)\, x_0 \tag{3.11}$$

where $x_0 = [x_{i0}]$ is an arbitrary initial vector of the dynamical system $x_0 = x(0)$.

Now the inhomogeneous equation

$$\dot{x}(t) = A x(t) + f(t)\,, \quad x(0) = x_0 \tag{3.12}$$

is considered. The general solution of (3.12) is given by superposition of the general solution (3.11) of the homogeneous system (3.2) and of a particular integral of (3.12). This particular solution is constructed by the method of variation of constants. This result in the unique general solution of (3.12)

$$x(t) = \Phi(t)\, x_0 + \int_0^t \Phi(t-\tau)\, f(\tau)\, d\tau \tag{3.13}$$

The state transition matrix is the fundamental concept for the solution of linear differential equations. But the solution of linear time-invariant dynamiçal systems can be characterized algebraically by eigenvalues and eigenvectors, too. They will lead to a second computation algorithm for the fundamental matrix.

3.2 Eigenvalues, Eigenvectors, Normal Coordinates

Looking for invariant modes of the homogeneous system (3.2) the statement

$$x(t) = \tilde{x}\, e^{\lambda t} \tag{3.14}$$

leads to the eigenvalue-eigenvector problem

$$(\lambda \mathbf{E} - \mathbf{A})\tilde{x} = 0. \tag{3.15}$$

For a nontrivial solution of (3.15) the characteristic matrix $(\lambda \mathbf{E} - \mathbf{A})$ has to be singular or λ has to be a zero of the characteristic polynomial of $\mathbf{A}$,

$$p(\lambda) = \det(\lambda \mathbf{E} - \mathbf{A}) = \lambda^n + a_1\lambda^{n-1} + \ldots + a_{n-1}\lambda + a_n = 0. \tag{3.16}$$

Such zeros $\lambda = \lambda_i$, $i = 1(1)n$, are called the eigenvalues of $\mathbf{A}$, and the corresponding nontrivial solution vectors $\tilde{x} = x_i$ are (right) eigenvectors of $\mathbf{A}$. (Left eigenvectors arise in the problem $\tilde{x}^T(\lambda \mathbf{E} - \mathbf{A}) = 0$).

The coefficients a_i of the characteristic polynomial associated with a real matrix $\mathbf{A}$ are real values. In particular, the relations

$$a_1 = -\mathrm{tr}\,\mathbf{A}, \quad a_n = (-1)^n \det \mathbf{A} \tag{3.17}$$

hold. Furthermore, if a complex eigenvalue λ_k exists,

$$\lambda_K = -\delta_K + i\omega_K \quad , \tag{3.18}$$

the conjugate complex value $\bar{\lambda}_K = -\delta_K - i\omega_K$ is also an eigenvalue. Likewise, corresponding eigenvectors are complex and conjugate complex, respectively.

Eigenvectors associated with distinct eigenvalues are linearly independent. In the case of multiple eigenvalues one has to distinguish between the multiplicity μ_i of the root λ_i of (3.16) and of the nullity ν_i, i.e. the number of the linearly independent solutions of (3.15), $\nu_i = n - \text{rank}(\lambda_i \mathbf{E} - \mathbf{A})$. The multiplicity of an eigenvalue is not less than its nullity $1 \leq \nu_i \leq \mu_i$.

a) Simple matrices: $\nu_i = \mu_i$ for all λ_i .

For each eigenvalue λ_i there exist μ_i linearly independent eigenvectors $\mathbf{x}_i$. Arranging all eigenvectors as columns of a (regular) modal matrix

$$\mathbf{X} = \left[\mathbf{x}_1 \,|\, \mathbf{x}_2 \,|\, \cdots \,|\, \mathbf{x}_n\right] \tag{3.19}$$

and defining a diagonal matrix $\mathbf{\Lambda}$ of the eigenvalues λ_i ,

$$\mathbf{\Lambda} = \mathbf{diag}\left[\lambda_1 \, , \, \ldots \, , \lambda_n\right] \, , \tag{3.20}$$

the complete solution of the eigen-problem (3.15) is compactly written

$$\mathbf{AX} = \mathbf{X\Lambda} \tag{3.21}$$

which is equivalent to the similarity transformation of $\mathbf{A}$ by $\mathbf{X}$:

(3.22) $$\mathbf{X}^{-1}\mathbf{A}\mathbf{X} = \mathbf{\Lambda}.$$

The linear modal transformation

(3.23) $$\mathbf{x}(t) = \mathbf{X}\,\bar{\mathbf{x}}(t)$$

applied to the homogeneous system (3.2) results in a decoupled representation of the dynamical system

(3.24a) $$\dot{\bar{\mathbf{x}}}(t) = \mathbf{\Lambda}\bar{\mathbf{x}}(t)$$

or

(3.24b) $$\dot{\bar{x}}_i(t) = \lambda_i \bar{x}_i(t)\,, \quad i = 1(1)n.$$

Therefore, the eigenvectors form a basis of so-called normal coordinates of (3.2).
Usually, most of the eigenvalues λ_i in (3.24b) will be complex. For physical interpretation of the results a real analogue of (3.24b) is prefered. For this purpose, the decoupled complex differential equations (3.24b) for each pair of complex and conjugate complex eigenvalues can be transformed into a real differential equation of second order. From $\dot{\bar{x}}_K = (-\delta_K + i\omega_K)\bar{x}_K$ and $\dot{\bar{x}}_{K+1} = (-\delta_K - i\omega_K)\bar{x}_{K+1}$ it follows with

(3.25) $$\begin{aligned} \bar{\bar{x}}_K &= \bar{x}_K + \bar{x}_{K+1}\,, \\ \bar{\bar{x}}_{K+1} &= i(\bar{x}_K - \bar{x}_{K+1}) \end{aligned}$$

the differential equations

$$\dot{\bar{\bar{x}}}_K = -\delta_K \bar{\bar{x}}_K + \omega_K \bar{\bar{x}}_{K+1} ,$$
$$\dot{\bar{\bar{x}}}_{K+1} = -\omega_K \bar{\bar{x}}_K - \delta_K \bar{\bar{x}}_{K+1} , \tag{3.26a}$$

or equivalently

$$\ddot{\bar{\bar{x}}}_K + 2\delta_K \dot{\bar{\bar{x}}}_K + \left(\delta_K^2 + \omega_K^2\right) \bar{\bar{x}}_K = 0 . \tag{3.26b}$$

While the transformation (3.23) of $\bar{\mathbf{x}}$ to $\mathbf{x}$ is generated in the case of complex eigenvalues by the complex eigenvectors $\mathbf{x}_K = \mathbf{x}_{KR} + \mathbf{x}_{KI}, \mathbf{x}_{K+1} = \mathbf{x}_{KR} - i\mathbf{x}_{KI}$, the transformation

$$\mathbf{x}(t) = \bar{\mathbf{X}}\, \bar{\bar{\mathbf{x}}}(t) \tag{3.27}$$

of $\bar{\bar{\mathbf{x}}}$ to $\mathbf{x}$ is generated by the real column vectors $\mathbf{x}_{KR}$ and $\mathbf{x}_{KI}$ summarized by the real matrix $\bar{\mathbf{X}}$.

b) <u>Defective matrices: $\nu_i < \mu_i$ for at least one eigenvalue: λ_i</u>

For at least one eigenvalue λ_i there exist only $\nu_i < \mu_i$ linearly independent eigenvectors

$$\mathbf{x}_{i_1} , \; \dots \; , \mathbf{x}_{i_{\nu_i}} .$$

To get a modal transformation similar to (3.19) the missing $\mu_i - \nu_i$ linearly independent vectors are determined by recursive linear equations

$$\left(\lambda_i \mathbf{E} - \mathbf{A}\right) \mathbf{x}_{i_K}^{(j)} = -\mathbf{x}_{i_K}^{(j-1)} , \tag{3.28}$$

$$j = 2, \ldots, \varrho \qquad x_{i_K}^{(1)} = x_{i_K}, \quad K = 1(1)\nu_i$$

breaking off for $j = \varrho_K$ since $j = \varrho_K + 1$ does not yield a solution vector. These generalized eigenvectors $x_{i_K}^{(j)}$ supplement the usual eigenvectors x_i to a complete basis of an eigenspace associated with the eigenvalue λ_i. Then, the generalized modal matrix $\mathbf{X}$ is given by s block matrices related to the s distinct eigenvalues λ_i :

(3.29)
$$\mathbf{X} = \left[\mathbf{X}_1 \,|\, \ldots \,|\, \mathbf{X}_s \right]$$

$$\mathbf{X}_i = \left[\mathbf{X}_{i_1} \,|\, \ldots \,|\, \mathbf{X}_{i_{\nu_i}} \right] = \left[x_{i_1} \,|\, x_{i_1}^{(2)} \,|\, \ldots \,|\, x_{i_1}^{(\varrho_1)} \,|\, \ldots \,|\, x_{i_{\nu_i}} \,|\, x_{i_{\nu_i}}^{(2)} \,|\, \ldots \,|\, x_{i_{\nu_i}}^{(\varrho_{\nu_i})} \right].$$
(3.30)

The corresponding similarity transformation represents the system matrix $\mathbf{A}$ in the Jordan canonical form:

(3.31)
$$\mathbf{X}^{-1}\mathbf{A}\mathbf{X} = \mathbf{J} = \operatorname{diag}\left[\ldots ; \mathbf{J}_{i_1}, \ldots, \mathbf{J}_{i_{\nu_i}} ; \ldots \right]$$

where the block diagonal elements are

(3.32)
$$\mathbf{J}_{i_K} = \begin{bmatrix} \lambda_i & 1 & 0 & \cdots & 0 \\ & \ddots & \ddots & \ddots & \vdots \\ & & \ddots & \ddots & 0 \\ & & & \ddots & 1 \\ & 0 & & & \lambda_i \end{bmatrix}_{(\varrho_K \times \varrho_K)}$$

In contrast to simple matrices (Case a) the Jordan canonical form of defective matrices is not a diagonal matrix but a matrix consisting of the λ_i 's in the diagonal and certain "1" in the first upper off-diagonal.
Further details will be found in the books of Zurmühl [1964] and Lancaster [1969] .

3.3 Solution by Eigenvalues and Eigenvectors

The similarity transformation $\mathbf{x}(t)=\mathbf{X}\,\bar{\mathbf{x}}(t)$ with the modal matrix (3.29) transfers the representation (3.2) of a linear time-invariant dynamical system in its Jordan canonical representation

$$\dot{\bar{\mathbf{x}}}(t) = \mathfrak{J}\,\bar{\mathbf{x}}(t), \quad \bar{\mathbf{x}}(0) = \mathbf{X}^{-1}\mathbf{x}(0) . \tag{3.33}$$

The state transition matrix of this system is given by

$$\bar{\Phi}(t) = e^{\mathfrak{J}t} = \mathbf{diag}\left[\ldots, e^{\mathfrak{J}_{i_1}t}, \ldots, e^{\mathfrak{J}_{i_{\nu_i}}t}; \ldots\right] \tag{3.34}$$

where

$$e^{\mathfrak{J}_{i_K}t} = e^{\lambda_i t}\begin{bmatrix} 1 & \frac{t}{1!} & \frac{t^2}{2!} & \cdots & \frac{t^{\varrho_K-1}}{(\varrho_K-1)!} \\ & 1 & \frac{t}{1!} & \ddots & \vdots \\ & & 1 & \ddots & \frac{t^2}{2!} \\ & 0 & & \ddots & \frac{t}{1!} \\ & & & & 1 \end{bmatrix}_{(\varrho_K \times \varrho_K)} . \tag{3.35}$$

Backtransformation to the original system (3.2) yields

$$\Phi(t) = X e^{Jt} X^{-1}. \tag{3.36}$$

In the case of simple matrices, $J = \Lambda$, the expression (3.36) of $\Phi(t)$ reduces

$$\Phi(t) = X e^{\Lambda t} X^{-1}, \quad e^{\Lambda t} = \operatorname{diag}\left(e^{\lambda_i t}\right). \tag{3.37}$$

Then, the general solution of the homogeneous differential equation (3.2) may be written in terms of eigenvalues and eigenvectors,

$$x(t) = \sum_{i=1}^{n} c_i x_i e^{\lambda_i t}, \tag{3.38}$$

where the coefficients c_i are related to the initial condition $x(0) = x_0$ by

$$c = [c_i] = X^{-1} x_0. \tag{3.39}$$

In general, the calculation of the solution $x(t)$ or equivalently of the fundamental matrix $\Phi(t)$ is reduced to the calculation of the elementary solutions $X_{i_K} e^{J_{iK} t}$ characterizing the eigenmodes of the system. The general solution is a linear combination of the elementary solutions represented in normal coordinates. Therefore, the dynamical behaviour of system (3.2) is completely characterized by the state transition matrix $\Phi(t)$ as shown in (3.11) or equivalently by the eigenvalues and (generalized) eigenvectors as shown e.g. in (3.38). Both concepts are connected by (3.36).

CHAPTER IV

Boundedness and Stability

The general solution of a vibration system has been found in chapter 3 for a given initial condition and a given forcing function. In technical applications, however, initial condition and forcing function are sometimes not exactly known, they are only specified as elements of a set of possible initial conditions or a family of forcing functions. Then, the behavior of the vibration system can be characterized by the qualitative properties of boundedness and stability.

Although most people have an intuitive feeling as to what stability means, the concept is very subtle, and rigorous definitions are necessary. They are discussed in section 4.1. This chapter is based on references by Cesari (1963), Lehnigk (1966), Müller (1974), Willems (1970).

4.1 Definitions

The concept of stability for general time-dependent non-linear systems is very complex. A very large number of definitions exists; only the most useful ones will be discussed in this section. Furthermore, this discussion will be restricted to finite-dimensional linear systems with constant system matrices,

$$\dot{x}(t) = \mathbf{A}\,x(t) + f(t), \quad x(0) = x_0 . \tag{4.1}$$

Basically the definitions can be divided in three classes. The first class of stability definitions deals with the response of forced systems to various inputs $f(t)$. The second class concerns the boundedness behavior of forced systems with a given input $f(t)$. The third class of stability definitions is related to the motions of free systems with respect to the initial conditions.

Definition 4.1: BIBO stability

The dynamical system (4.1) is called bounded-input-bounded-output (BIBO) stable if <u>any</u> bounded input $f(t)$ produces a bounded output $x(t)$, regardless of the bounded initial state x_0.

Definition 4.2: Boundedness (stability in the sense of Lagrange)

The dynamical system (4.1) is called bounded or Lagrange stable, with respect to a <u>given</u> input $f(t)$, if the output $x(t)$ is bounded, regardless of the (bounded) initial state x_0

Definition 4.3: Stability in the sense of Lyapunov

The equilibrium state $\mathbf{x} = \mathbf{0}$ of the free system

$$\dot{\mathbf{x}}(t) = \mathbf{A}\mathbf{x}(t), \quad \mathbf{x}(0) = \mathbf{x}_0 \tag{4.2}$$

is called stable (in the sense of Lyapunov) if for any positive ε there exists a positive $\delta = \delta(\varepsilon)$ such that

$$\| \mathbf{x}_0 \| < \delta \tag{4.3}$$

implies

$$\| \mathbf{x}(t) \| < \varepsilon \tag{4.4}$$

for all $t \geq 0$.

Definition 4.4: Asymptotic stability

The equilibrium state $\mathbf{x} = \mathbf{0}$ of the free system (4.2) is called asymptotically stable if it is stable and

$$\lim_{t \to \infty} \mathbf{x}(t) = \mathbf{0} . \tag{4.5}$$

In engineering language, the concept of stability in the sense of Lyapunov can be interpreted as follows: if a system is displaced from the eqiulibrium state, then its motion remains in a corresponding neighborhood of the equilibrium state. Asymptotic stability is stronger: it requires in addition that the motion returns to the equilibrium after perturbation.

The definitions 4.3 and 4.4 are not concerned with properties of systems, but of equilibrium solutions of the system equation. However, in the case of linear systems (4.2), the stability of each motion can be reduced to the stability of the equilibrium solution. Hence, linear systems can be classified as stable or asymptotically stable systems depending on the stability of the equilibrium state.

The free system (4.2) is said to be unstable if it is not stable.

For technical reasons not only stability with respect to perturbations in the initial conditions is of interest but even more the behavior of the system with respect to external disturbances. Therefore, BIBO stability has to be investigated too. For linear systems some very interesting results have been obtained on the relationship between different types of stability. These are discussed in the next section.

4.2 Criteria of Boundedness and of BIBO Stability

Obviously, BIBO stability implies boundedness and asymptotic stability implies stability. More interesting is the relationship between asymptotic stability and BIBO stability.

Theorem 4.1: BIBO and asymptotic stability (I)

The forced linear dynamical system (4.1) is BIBO stable if and only if the free linear dynamical system (4.2) is asymptotically stable.

The exact proof is omitted, but following considerations may be helpful for a plausible explanation of theorem 4.1. If the system (4.1) is BIBO stable, the system (4.2) must be at least stable for $\mathbf{f} = \mathbf{0}$. But if (4.2) is only stable a periodic bounded vector function $\mathbf{f}(t)$ always exists such that the motion $\mathbf{x}(t)$ of (4.2) will be unbounded (see section 6.2). Therefore, the system must be asymptotically stable. In the opposite, if the free system is asymptotically stable, the fundamental matrix $\boldsymbol{\Phi}(t)$ is exponentially bounded,

$$\|\boldsymbol{\Phi}(t)\| < c_1 e^{-c_2 t} , \quad c_i > 0 , \quad i = 1,2 ,$$

(see section 4.3.1) and the general solution of (4.1), given by equation (3.13), satisfies the inequality

$$\|\mathbf{x}(t)\| \leq \|\boldsymbol{\Phi}(t)\| \, \|\mathbf{x}_0\| + \int_0^t \|\boldsymbol{\Phi}(t-\tau)\| \, \|\mathbf{f}(\tau)\| \, d\tau$$

$$\leq c_1 \|\mathbf{x}_0\| + c_1 c_3 \int_0^t e^{-c_2(t-\tau)} \, d\tau$$

$$\leq c_1 \|\mathbf{x}_0\| + \frac{c_1 c_3}{c_2}$$

where $\| f(t) \| < c_3$: a bounded input produces a bounded output. The forced system is BIBO stable.

Often the forcing vector function $f(t)$ is restricted to $r\,(<n)$ functions $u_i(t)$:

(4.6) $$f(t) = B\,u(t)$$

where $u(t) = [u_i(t)]$ is a r-dimensional vector and B a $n \times r$-matrix, e.g. for mechanical systems $f(t) = [0 | M^{-1}]^T h(t)$. Then theorem 4.1 is not valid. But a slight modification results in a similar theorem.

Theorem 4.2: BIBO and asymptotic stability (II)

Asymptotic stability of (4.2) implies BIBO stability of the dynamical system

(4.7) $$\dot{x}(t) = A\,x(t) + B\,u(t)\,, \quad x(0) = x_0\,.$$

BIBO stability of (4.7) implies stability of (4.2). Moreover, if (4.7) is completely controllable, i.e.

(4.8) $$\text{rank}\left[B \,|\, AB \,|\, A^2 B \,|\, \ldots \,|\, A^{n-1} B \right] = n$$

(see Appendix A), BIBO stability of (4.7) implies asymptotic stability of (4.2).

Finally, if the forcing vector function is restricted to a given $f(t)$ two relations of stability and boundedness can be established.

Theorem 4.3: Boundedness and asymptotic stability

Asymptotic stability of (4.2) implies boundedness of (4.1).

This theorem follows obviously by the definitions and by theorem 4.1.

Theorem 4.4: Boundedness and stability

Boundedness of (4.1) implies stability of (4.2).

Vice versa, stability does not imply boundedness; this depends on the given type of forcing function: (i) In the case of a bounded periodic function $f(t)=f(t+T)$ stability of (4.2) implies boundedness of (4.1) if and only if the frequencies given by the purely imaginary eigenvalues of $\mathbf{A}$ do not coincide with the frequencies of the nonvanishing Fourier terms of $f(t)$; (ii) in the case of forcing functions bounded by $\int_0^\infty \|f(t)\|\, dt < \infty$ stability of (4.2) implies boundedness of (4.1).

The theorems 4.1-4.4 justify to investigate only stability and asymptotic stability in more detail which will be done in the next section.

4.3 Criteria of Stability and Asymptotic Stability

This section deals with the stability and asymptotic stability of linear time-invariant dynamic systems (4.2). Necessary and sufficient conditions for asymptotic stability have already been known for a century; the early work of Hermite was published in 1850. The best known criteria were found by Routh in 1877 and Hurwitz in 1895. Another class of stability criteria is due to Lyapunov 1892, presenting the mathematical background of stability criteria for mechanical systems due to Thomson and Tait, 1879. This second class of criteria has today a revival in modern system theory summarized e.g. by Müller (1974).

4.3.1 Stability and Eigenvalues

The stability of linear systems (4.2) is invariant to linear coordinate transformations $\mathbf{x}(t) = \mathbf{T}\bar{\mathbf{x}}(t)$, $\det \mathbf{T} \neq 0$. The requirements (4.3) - (4.5) on $\mathbf{x}(t)$ are satisfied if and only if (4.3) - (4.5) are valid for $\bar{\mathbf{x}}(t)$, too. Therefore, the dynamical system is investigated in the Jordan canonical representation (3.32) instead of (4.2). From the solution

$$\bar{\mathbf{x}}(t) = e^{\mathbf{J}t}\,\bar{\mathbf{x}}_0 \tag{4.9}$$

the following stability theorem holds.

Theorem 4.5: Stability behavior dependent on eigenvalues

The linear time-invariant dynamical system (4.2) is

asymptotically stable, if all the eigenvalues of the system matrix $\mathbf{A}$ have negative real parts: $\mathrm{Re}\,\lambda_i < 0$, $i = 1(1)n$;

marginally stable, if all the eigenvalues have non positive real parts and some of them actually have a zero real part where the multiplicities μ_i and the nullities ν_i of these eigenvalues with vanishing real parts are equal, respectively $\mathrm{Re}\,\lambda_i \leq 0$, $i = 1(1)n$ and $\mathrm{Re}\,\lambda_j = 0 : \nu_j = \mu_j$;

unstable, if at least one of the eigenvalues has a positive real part, or if there exists at least one eigenvalue with vanishing real part where the nullity is less than the multiplicity: $\mathrm{Re}\,\lambda_i > 0$ or $\mathrm{Re}\,\lambda_i = 0$ and $\nu_i < \mu_i$.

The criterion for asymptotic stability $\mathrm{Re}\,\lambda_i < 0$, $i = 1(1)n$, follows immediately by the system representation (4.9) with $e^{\mathbf{J}t}$ given by (3.33) and (3.34). Likewise, instability arises if at least one eigenvalue has a positive real part. Only the critical case of eigenvalues with vanishing real part has to be considered in more detail. The corresponding elementary solution is bounded if and only if no time-weighted exponential $\left(t^{\kappa}e^{\lambda_i t}\right)$ exists. This is valid if and only if $\nu_i = \mu_i$ is valid for these critical eigenvalues. From this fact the statements of stability and instability in theorem 4.5 follow.

Theorem 4.5 characterizes the stability of (4.2) by the eigenvalues of $\mathbf{A}$. Therefore the dynamical problem of stability is reduced to the algebraic problem of eigenvalue distribution of a matrix. This algebraic problem can be solved by two different approaches: the Routh-Hurwitz approach by the characteristic polynomial and the Lyapunov approach by the so-called Lyapunov matrix equation.

4.3.2 Routh-Hurwitz Criteria

The eigenvalues λ_i of $\mathbf{A}$ are the zeros of the characteristic polynomial (3.16). Therefore the eigenvalues are completely determined by the characteristic coefficients a_i , $i = 1(1)n$. It can be found conditions for the a_i's to guarantee that all λ_i have negative real parts. They are given by theorems due to Routh, Hurwitz and Liénard and Chipart.

Theorem 4.6: Necessary condition for asymptotic stability (Stodola condition)

A necessary condition that all zeros of the real

characteristic polynomial (3.16) have negative real parts (i.e. that system (4.2) is asymptotically stable) is that all coefficients $a_i \; (i = 1(1)n)$ are positive :

$$a_i > 0 \;, \quad i = 1(1)n \,. \tag{4.10}$$

Theorem 4.7: Routh criterion

All zeros of the real characteristic polynomial (3.16) have negative real parts, i.e. the system (4.2) is asymptotically stable, if and only if all Routh numbers $R_i \; (i = 1(1)n)$ are positive :

$$R_i > 0 \;, \quad i = 1(1)n \,. \tag{4.11}$$

The Routh numbers R_i are given by the elements of the first column in the Routh array associated with $p(\lambda)$ (3.16).

<u>Routh-array</u>

$$\begin{array}{c|l|lll}
 & \qquad\quad 1 & a_2 & a_4 & a_6 \\
 & R_1 = \quad a_1 & a_3 & a_5 & a_7 \\
\hline
r_2 = \dfrac{1}{R_1} & R_2 = c_{21} = a_2 - r_2 a_3 & c_{22} = a_4 - r_2 a_5 & c_{23} = a_6 - r_2 a_7 & \text{--------} \\
r_3 = \dfrac{R_1}{R_2} & R_3 = c_{31} = a_3 - r_3 c_{22} & c_{32} = a_5 - r_3 c_{23} & \text{---------} & \\
 & \vdots & & & \\
r_j = \dfrac{R_{j-2}}{R_{j-1}} & R_j = c_{j1} & c_{jk} = c_{j-2,k+1} - r_j c_{j-1,k+1} & & \\
 & \vdots & & & \\
 & R_n = a_n & & &
\end{array} \tag{4.12}$$

Theorem 4.8: Hurwitz criterion

All zeros of the real characteristic polynomial (3.16) have negative real parts, i.e. the system (4.2) is asymptotically stable, if and only if all Hurwitz determinants H_i $(i=1(1)n)$ are positive :

(4.13) $$H_i > 0 \ , \quad i = 1\,(1)\,n \, .$$

The Hurwitz determinants H_i are the main principal minors of the Hurwitz matrix associated with the characteristic polynomial (3.16)

(4.14) $$\mathbf{H} = \begin{bmatrix} a_1 & 1 & 0 & 0 & \cdots & 0 \\ a_3 & a_2 & a_1 & 1 & & \vdots \\ a_5 & a_4 & a_3 & a_2 & & \vdots \\ a_7 & a_6 & a_5 & a_4 & & \vdots \\ \vdots & \vdots & a_7 & a_6 & & \vdots \\ \vdots & \vdots & \vdots & \vdots & & \vdots \\ 0 & 0 & 0 & 0 & \cdots & a_n \end{bmatrix} ,$$

$$H_1 = a_1 \ , \quad H_2 = a_1 a_2 - a_3 \ , \ \ldots\ldots , \ H_n = a_n H_{n-1} = \det \mathbf{H} \, .$$

Combining the necessary conditions (4.10) and the Hurwitz conditions (4.13) a more simple theorem is obtained.

Theorem 4.9: Liénard-Chipart criterion

All zeros of the real characteristic polynomial (3.16) have negative real parts, i.e. the system (4.2) is asimptotically stable, if and only if the n conditions

$$a_n > 0 \ , \ H_{n-1} > 0 \ , \ a_{n-2} > 0 \ , \\ H_{n-3} > 0 \ , \ \ldots\ldots \ , \ H_1 = a_1 > 0 \tag{4.16}$$

are satisfied.

Remark: Theorems 4.6-4.9 allow to check asymptotic stability. But there is no equivalent theorem to check only stability.

4.3.3 Lyapunov Criteria

Although Lyapunov's stability theory is very famous for general nonlinear time-variable dynamical systems, his algebraic criteria to the linear time-invariant stability problem are not as well-known. But today, in modern system and control theory it is important to be familiar with his results and their extensions.

For a plausibility interpretation consider firstly the square root of a positive definite quadratic form as a special vector norm

$$\|\mathbf{x}(t)\|^2 = \mathbf{x}^T(t)\mathbf{R}\,\mathbf{x}(t) \ ; \quad \mathbf{R} = \mathbf{R}^T > 0 \ . \tag{4.17}$$

Then, the time derivative of (4.17) along a trajectory of (4.2) leads to

$$\frac{d}{dt}\|\mathbf{x}(t)\|^2 = \mathbf{x}^T(t)\,(\mathbf{A}^T\mathbf{R} + \mathbf{R}\mathbf{A})\,\mathbf{x}(t). \tag{4.18}$$

If it is possible, secondly, to keep constant or to reduce $\|\mathbf{x}(t)\|^2$ along any trajectory of (4.2), i.e.

$$\frac{d}{dt}\|\mathbf{x}(t)\|^2 \stackrel{!}{=} -\mathbf{x}^T(t)\mathbf{S}\mathbf{x}(t), \quad \mathbf{S} = \mathbf{S}^T \geq \mathbf{0}, \tag{4.19}$$

where $\mathbf{S}$ is a nonnegative definite matrix, then the linear system (4.2) must be stable or asymptotically stable by the definitions 4.3 and 4.4. Equations (4.18) and (4.19) have to be satisfied simultaneously for all state vectors $\mathbf{x}$. Therefore, $\mathbf{R}$ and $\mathbf{S}$ are related by the Lyapunov matrix equation

$$\mathbf{A}^T\mathbf{R} + \mathbf{R}\mathbf{A} = -\mathbf{S}. \tag{4.20}$$

Similar to section 4.3.2 the characteristic coefficients are determining the stability behavior of (4.2). Here the solution $\mathbf{R}$ of (4.20) with given $\mathbf{S} = \mathbf{S}^T \geq \mathbf{0}$ characterizes completely the eigenvalue distribution of $\mathbf{A}$.

In Appendix B the properties of the Lyapunov equation are reviewed; here we are only interested in the stability results.

Theorem 4.10: Lyapunov criterion

The linear time-invariant system (4.2) is asymptotically stable if and only if for any given symmetric, posi-

tive definite matrix **S** there exists a symmetric, positive definite matrix **R** which is the unique solution of the Lyapunov matrix equation (4.20).
To check asymptotic stability by this theorem it suffices to solve (4.20) only for one given $\mathbf{S} = \mathbf{S}^T > 0$ (e.g. $\mathbf{S} = \mathbf{E}$) because if (4.2) is asymptotically stable then each $\mathbf{S} = \mathbf{S}^T > 0$ leads uniquely to a symmetric, positive definite solution matrix **R** of (4.20).

Recently, some useful extensions of the Lyapunov stability theorem were developed which are of particular interest for mechanical systems, Müller (1974).

Theorem 4.11: Asymptotic stability theorem

The linear time-invariant system (4.2) is asymptotically stable if and only if there exists a unique, symmetric, positive definite solution matrix **R** of (4.20) for at least one (and hence for any) given symmetric, positive semidefinite matrix **S** which satisfies the observability condition

$$\operatorname{rank}\left[\mathbf{S} \,|\, \mathbf{A}^T\mathbf{S} \,|\, \ldots\ldots \,|\, \mathbf{A}^{T^{n-1}}\mathbf{S}\right] = n \qquad (4.21)$$

with respect to the system matrix **A**. (The meaning of observability is discussed in Appendix A).

Obviously, theorem 4.10 is a special case of theorem 4.11: for a positive definite matrix **S** the condition (4.21) is trivially satisfied.

Theorem 4.12: Stability theorem

The linear time-invariant system (4.2) is marginally stable if and only if there exists a (not necessarily unique) symmetric, positive definite solution matrix $\mathbf{R}$ of (4.20) for at least one symmetric, positive, semidefinite matrix $\mathbf{S}$ which violates the observability condition (4.21).

Theorem 4.13: Instability theorem

The linear time-invariant system (4.2) is unstable if and only if there exists a symmetric, positive semidefinite matrix $\mathbf{S}$ which leads to a solution matrix $\mathbf{R}$ with the property of $\mathbf{x}^T\mathbf{R}\mathbf{x} < 0$ for an arbitrary observable state vector $\mathbf{x}$,

$$\mathbf{x} = \left[\mathbf{S} \mid \mathbf{A}^T\mathbf{S} \mid \ldots\ldots \mid \mathbf{A}^{T^{n-1}}\mathbf{S}\right]\mathbf{z}\,, \tag{4.22}$$

where $\mathbf{z}$ is a suitable $n^2 \times 1$ vector.

Especially, theorem 4.13 contains Lyapunov's sufficient instability theorem choosing $\mathbf{S}$ as a positive definite matrix.

As an application of theorem 4.11-4.13 let consider the stability problem of the following linear autonomous discrete mechanical system

$$\mathbf{M}\ddot{\mathbf{y}} + (\mathbf{D}+\mathbf{G})\dot{\mathbf{y}} + \mathbf{K}\mathbf{y} = \mathbf{0} \tag{4.23}$$

where $\mathbf{y}$ means the $f \times 1$ vector of generalized coordinates,

$\mathbf{M} = \mathbf{M}^T > 0$ the mass matrix, $\mathbf{D} = \mathbf{D}^T \geq 0$ the damping matrix of dissipative forces, $\mathbf{G} = -\mathbf{G}^T$ the matrix of gyroscopic forces, $\mathbf{K} = \mathbf{K}^T$ ($\det \mathbf{K} \neq 0$) the nonsingular spring matrix of forces derivable from a potential (see also Chapter 2). The state space representation of (4.23) was given in section 2.4 by

$$\dot{\mathbf{x}} = \underbrace{\begin{bmatrix} \mathbf{0} & \mathbf{E} \\ -\mathbf{M}^{-1}\mathbf{K} & -\mathbf{M}^{-1}(\mathbf{D}+\mathbf{G}) \end{bmatrix}}_{\mathbf{A}} \mathbf{x}\,, \quad \mathbf{x} = \begin{bmatrix} \mathbf{y} \\ \dot{\mathbf{y}} \end{bmatrix}, \quad n = 2f\,. \qquad (4.24)$$

Investigating the stability of (4.23) or equivalently of (4.24) the Hamiltonian of (4.23) is chosen as a quadratic form for (4.17)

$$H = \frac{1}{2}\left(\dot{\mathbf{y}}^T \mathbf{M} \dot{\mathbf{y}} + \mathbf{y}^T \mathbf{K} \mathbf{y}\right) = \frac{1}{2}\mathbf{x}^T \begin{bmatrix} \mathbf{K} & \mathbf{0} \\ \mathbf{0} & \mathbf{M} \end{bmatrix} \mathbf{x} = \mathbf{x}^T \mathbf{R} \mathbf{x}\,. \qquad (4.25)$$

The time derivative of (4.25) is written as

$$\dot{H} = -\dot{\mathbf{y}}^T \mathbf{D} \dot{\mathbf{y}} = -\mathbf{x}^T \begin{bmatrix} \mathbf{0} & \mathbf{0} \\ \mathbf{0} & \mathbf{D} \end{bmatrix} \mathbf{x} = -\mathbf{x}^T \mathbf{S} \mathbf{x}\,. \qquad (4.26)$$

From this the following stability theorem is obvious.

Theorem 4.14: Thomson-Tait-Chetaev stability theorem with semidefinite damping

When the Hamiltonian (4.25) of the mechanical

system (4.23) is positive definite, i.e. if $\mathbf{K} = \mathbf{K}^T > \mathbf{0}$, the system is at least stable. Moreover, if the damping is pervasive, i.e.

$$\text{rank}\left[\mathbf{S} \mid \mathbf{A}^T\mathbf{S} \mid \ldots\ldots \mid \mathbf{A}^{T^{n-1}}\mathbf{S}\right] = n \tag{4.21}$$

with $\mathbf{A}$ of (4.24) and $\mathbf{S}$ of (4.26), then the system is asymptotically stable if and only if $\mathbf{K} = \mathbf{K}^T > \mathbf{0}$, Theorem 4.11, and is unstable if $\mathbf{K} \not> \mathbf{0}$ $(\det \mathbf{K} \neq 0)$, Theorem 4.13.

This theorem was stated by Thomson and Tait in 1879 and by Chetaev in 1946 for definite damping $\mathbf{D} > \mathbf{0}$. The extension to semidefinite damping $\mathbf{D} \geq \mathbf{0}$ by means of a controllability or observability condition was given by Müller (1970).

Example 4.1 Automobile wheel suspension

In example 2.1 the problem of an automobile wheel suspension was considered. The equation of motion reads as

$$\begin{bmatrix} m_1 & 0 \\ 0 & m_2 \end{bmatrix} \ddot{\mathbf{y}} + \begin{bmatrix} d_1 & -d_1 \\ -d_1 & d_1 \end{bmatrix} \dot{\mathbf{y}} + \begin{bmatrix} c_1 & -c_1 \\ -c_1 & c_1+c_2 \end{bmatrix} \mathbf{y} = \mathbf{0}.$$

For the stability analysis of this system theorem 4.9 (Liénard-Chipart) and theorem 4.14 (Thomson-Tait-Chetaev) may be applied. The characteristic polynomial is

$$p(\lambda) = \lambda^4 + d_1\left(\frac{1}{m_1} + \frac{1}{m_2}\right)\lambda^3 + \left(\frac{c_1}{m_1} + \frac{c_1}{m_2} + \frac{c_2}{m_2}\right)\lambda^2 +$$

$$+ \frac{d_1}{m_1} \frac{c_2}{m_2} \lambda + \frac{c_1 c_2}{m_1 m_2} .$$

Here the Liénard-Chipart conditions are

$$a_4 = \frac{c_1 c_2}{m_1 m_2} > 0 ,$$

$$H_3 = a_1 a_2 a_3 - a_1^2 a_4 - a_3^2 = \frac{d_1^2 c_2^2}{m_1 m_2^3} > 0 ,$$

$$a_2 = c_1 \left(\frac{1}{m_1} + \frac{1}{m_2} \right) + \frac{c_2}{m_2} > 0 ,$$

$$H_1 = a_1 = d_1 \left(\frac{1}{m_1} + \frac{1}{m_2} \right) > 0$$

which yields the stability conditions

$$d_1 > 0 , \quad c_1 > 0 , \quad c_2 > 0 . \qquad (4.27)$$

Theorem 4.14 yields the conditions (4.27) immediately. Only the question of pervasive damping has to be considered in more detail. In the case of a vanishing gyroscopic matrix **G** the observability condition (4.21) of theorem 4.14 can be replaced by the controllability condition

$$\text{rank} \left[\mathbf{M}^{-1}\mathbf{D} \,\middle|\, (\mathbf{M}^{-1}\mathbf{K})\mathbf{M}^{-1}\mathbf{D} \,\middle|\, \ldots \,\middle|\, (\mathbf{M}^{-1}\mathbf{K})^{f-1}\mathbf{M}^{-1}\mathbf{D} \right] = f . \qquad (4.28)$$

Here one has to calculate

$$\text{rank} \left[\begin{array}{cc|cc} \frac{d_1}{m_1} & -\frac{d_1}{m_1} & \frac{c_1 d_1}{m_1}\left(\frac{1}{m_1}+\frac{1}{m_2}\right) & -\frac{c_1 d_1}{m_1}\left(\frac{1}{m_1}+\frac{1}{m_2}\right) \\ -\frac{d_1}{m_2} & \frac{d_1}{m_2} & -\frac{c_1 d_1}{m_2}\left(\frac{1}{m_1}+\frac{1}{m_2}\right)-\frac{c_2 d_1}{m_2^2} & \frac{c_1 d_1}{m_2}\left(\frac{1}{m_1}+\frac{1}{m_2}\right)+\frac{c_2 d_1}{m_2^2} \end{array} \right] =$$

$$= \operatorname{rank} \begin{bmatrix} \frac{d_1}{m_1} & -\frac{d_1}{m_1} & \frac{c_1 d_1}{m_1}\left(\frac{1}{m_1}+\frac{1}{m_2}\right) & -\frac{c_1 d_1}{m_1}\left(\frac{1}{m_1}+\frac{1}{m_2}\right) \\ 0 & 0 & -\frac{c_2 d_1}{m_2^2} & \frac{c_2 d_1}{m_2^2} \end{bmatrix} = 2 .$$

Although the damping force is only acting between the two masses m_1 and m_2 (see example 2.1) the system is pervasively damped and therefore asymptotically stable for $d_1 > 0$, $c_1 > 0$, $c_2 > 0$.

Example 4.2: Centrifuge

The motion of the centrifuge considered in example 2.2 is determined by the normalized matrix differential equation

$$\begin{bmatrix} \ddot{\phi} \\ \ddot{\theta} \end{bmatrix} + (\delta \mathbf{E} + \omega \mathbf{S}) \begin{bmatrix} \dot{\phi} \\ \dot{\theta} \end{bmatrix} + k \begin{bmatrix} \phi \\ \theta \end{bmatrix} = \mathbf{0} .$$

For $\delta > 0, k > 0$ the system is asymptotically stable by theorem 4.14. Also by the characteristic polynomial

$$p(\lambda) = \lambda^4 + 2\delta\lambda^3 + (2k + \delta^2 + \omega^2)\lambda^2 + 2\delta k\lambda + k^2 =$$

$$= [\lambda^2 + (\delta + i\omega)\lambda + k][\lambda^2 + (\delta - i\omega)\lambda + k]$$

asymptotic stability for $\delta > 0$, $k > 0$ can be easily shown via

Routh-Hurwitz conditions. The special case of $\delta = 0$ and $k < 0$ can be considered, too, due to the factorization of $p(\lambda)$:

$$\lambda_{1,2,3,4} = \pm \frac{i}{2}\left[\omega \pm \sqrt{\omega^2 + 4k}\right] \quad (\delta = 0).$$

The eigenvalues will be purely imaginary if and only if

$$\omega^2 > -4k \qquad (k < 0).$$

In this case the system is stabilized by gyroscopic forces although the spring forces are statically unstable. But this effect of gyroscopic stabilization is only possible if there is no damping. Therefore, in real systems with damping (e. g. by friction) gyroscopic stabilization is not possible.

CHAPTER 5

Special Responses of Linear Vibration Systems

Mechanical vibration systems are often excited by forces with special time history. Such special excitations are the impulse forces and the periodic forces. Impulse forces appear, in particular, at vehicle vibration systems, e.g. an automobile driving on a good road with some holes, an airplane flying through a turbulence or a satellite hitting a meteor. Periodic forces are found in vibration systems with unbalanced rotating parts, e.g. an unbalanced centrifuge or an automobile with unbalanced wheels. But periodic forces may be caused by a rough road or sea, too, affecting vehicles and ships. Further, in machine dynamics periodic forces are the primary disturbances. Therefore, an analysis of the system response due to impulse and periodic forces is well justified.

5.1 Excitation by Impulse Forces

The impulse forces may be modeled by the Dirac function $\delta(t-t_I)$:

$$\mathbf{f}(t) = \mathbf{f}_I \, \delta(t-t_I) \tag{5.1}$$

where $\mathbf{f}_I$ is a constant $n \times 1$ -vector and t_I is the time of the impulse. The Dirac function has the following properties

$$
\begin{aligned}
&\delta(t-t_I) = 0 && \text{for} \quad t < t_{I-}, \\
&\delta(t-t_I) \rightarrow \infty && \text{for} \quad t_{I+} \leq t \leq t_{I+}, \\
&\delta(t-t_I) = 0 && \text{for} \quad t > t_{I+}, \\
&\int_{t_{I-}}^{t_{I+}} \delta(t-t_I)\,dt = 1, && t_{I+} - t_{I-} \rightarrow 0,
\end{aligned}
\tag{5.2}
$$

where t_{I-} is the instant immediately before the impulse and t_{I+} is the instant immediately after the impulse, Fig. 5.1. Obviously the Dirac function is a mathematical idealization of a real impulse. However, this idealized model facilitates the solution considerably.

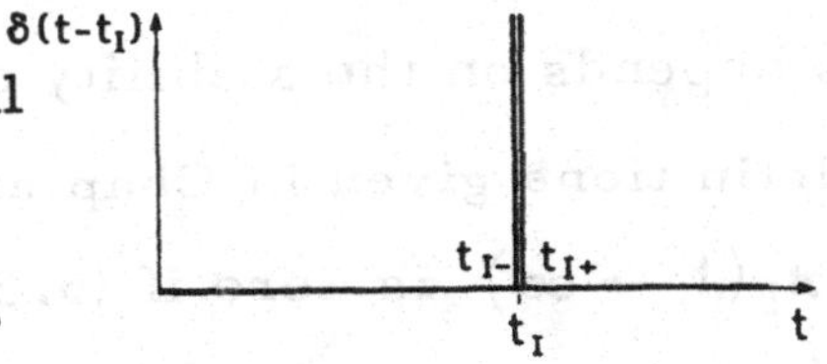

Fig.5.1. Dirac function

The general solution of the vibration system's differential equation

$$\dot{\mathbf{x}}(t) = \mathbf{A}\mathbf{x}(t) + \mathbf{f}(t) \tag{5.3}$$

is given by (3.13) as

$$\mathbf{x}(t) = \boldsymbol{\Phi}(t)\mathbf{x}_0 + \int_0^t \boldsymbol{\Phi}(t-\tau)\mathbf{f}(\tau)\,d\tau. \tag{5.4}$$

Introducing (5.1) in (5.4) one obtains the impulse response

$$\mathbf{x}_I(t) = \boldsymbol{\Phi}(t)\mathbf{x}_0 + \int_0^{t_{I-}} \boldsymbol{\Phi}(t-\tau)\mathbf{f}_I\delta(\tau-t_I)\,d\tau + \tag{5.5a}$$

$$(5.5b) \qquad + \int_{t_{I-}}^{t_{I+}} \Phi(t-\tau) f_I \delta(\tau - t_I) d\tau + \int_{t_{I+}}^{t} \Phi(t-\tau) f_I \delta(\tau - t_I) d\tau .$$

Using the properties (5.2), the impulse response can be represented as

$$(5.6) \qquad \begin{aligned} x_I(t) &= \Phi(t) x_0 && \text{for } t < t_{I-} , \\ x_I(t) &= \Phi(t)\left[x_0 + \Phi(-t_I) f_I\right] && \text{for } t > t_{I+} . \end{aligned}$$

This means that the impulse forces result in a singular variation of the initial vector at impulse time t_I .

The steady-state behavior to impulse excitations depends on the stability of the vibration system (5.3). From the definitions given in Chapter 4 it follows:

$x_I(t \rightarrow \infty)$ is zero if (5.3) asymptotically stable,
$x_I(t \rightarrow \infty)$ is bounded if (5.3) stable,
$x_I(t \rightarrow \infty)$ is unbounded if (5.3) unstable.

5.2 Excitation by Periodic Forces

The periodic forces may be given by

$$(5.7) \qquad f(t) = f(t+T)$$

where $f(t)$ is a $n \times 1$ -vector with period T . The periodic forces (5.7) can be expanded by a Fourier series

$$f(t) = \frac{1}{2} f_0 + \sum_{k=1}^{\infty} \left(f_k^{(1)} \cos k\Omega t + f_k^{(2)} \sin k\Omega t \right) \qquad (5.8)$$

where f_0 is a constant $n \times 1$ -vector and $f_k^{(1)}$, $f_k^{(2)}$ are $n \times 1$ -vectors of the Fourier coefficients. The period is given by

$$T = \frac{2\pi}{\Omega} \qquad (5.9)$$

where Ω is the frequency of the forces, Fig. 5.2.

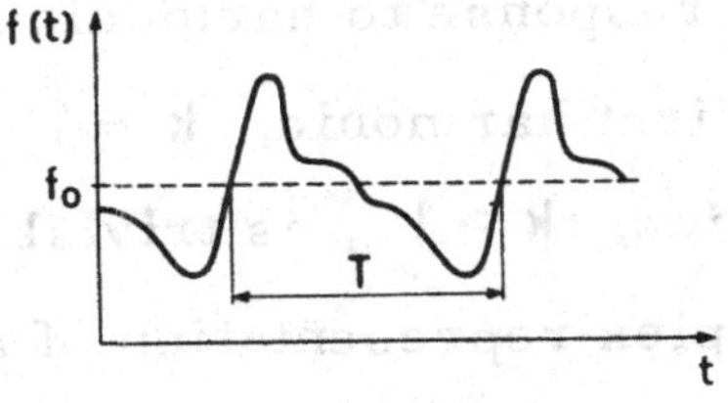

Fig.5.2. Periodic function

Introducing (5.8) in the general solution (5.4), one obtains the response to periodic forces

$$x_p(t) = \Phi(t) x_0 + \frac{1}{2} \int_0^t \Phi(t-\tau) f_0 \, d\tau + \sum_{k=1}^{\infty} \int_0^t \Phi(t-\tau) \left[f_k^{(1)} \cos k\Omega t + f_k^{(2)} \sin k\Omega t \right] d\tau \qquad (5.10)$$

where $\Phi(t-\tau) = e^{A(t-\tau)}$ according to (3.7) may be substituted. Thus, the response to periodic forces is available if the integrals in (5.10) are solved. There appear only two different kinds of integrals which will now be investigated in more detail.

The response to the constant force $f(t) = \frac{1}{2} f_0$ reads as

(5.11) $$x_{p0}(t) = \frac{1}{2}\int_0^t e^{A(t-\tau)} f_0 \, d\tau = (E - e^{At})\, x_\infty$$

where

(5.12) $$x_\infty = -\frac{1}{2} A^{-1} f_0$$

is a constant $n \times 1$ -vector. Thus, for the computation of x_∞ the system matrix **A** has to be regular.

The response to harmonic forces has only to be determined for the first harmonic, $k = 1$. Then, the result for higher harmonics, $k > 1$, is trivial. For a convenient computation, the complex representation of a harmonic function will be used:

(5.13) $$f(t) = f^{(1)}\cos\Omega t + f^{(2)}\sin\Omega t = f e^{i\Omega t} + \bar{f} e^{-i\Omega t}$$

where

(5.14) $$f = \frac{1}{2}\left(f^{(1)} - i f^{(2)}\right)$$

is a complex $n \times 1$ -vector. It has to be pointed out that the vector (5.13) of a harmonic force has components with different amplitudes e_i and phase angles φ_i :

$$f_i(t) = f_i^{(1)}\cos\Omega t + f_i^{(2)}\sin\Omega t = e_i \cos(\Omega t - \varphi_i),$$

(5.15)

$$e_i = \sqrt{f_i^{(1)2} + f_i^{(2)2}}\ , \quad \tan\varphi_i = \frac{f_i^{(2)}}{f_i^{(1)}}\ , \quad i = 1(1)n\ .$$

The response to the harmonic force (5.13) follows as

$$x_{p1}(t) = \int_0^t e^{A(t-\tau)}\left(f e^{i\Omega t} + \bar{f} e^{-i\Omega t}\right) d\tau =$$

$$= g e^{i\Omega t} + \bar{g} e^{-i\Omega t} - e^{At} g^{(1)} \qquad (5.16)$$

where

$$g = \frac{1}{2}\left(g^{(1)} - i g^{(2)}\right) = F f \qquad (5.17)$$

is a complex $n \times 1$ -vector and

$$F = \left(i\Omega E - A\right)^{-1} \qquad (5.18)$$

is the $n \times n$ -frequency response matrix. In (5.16), it appears a harmonic function

$$x_{p\infty}(t) = g e^{i\Omega t} + \bar{g} e^{-i\Omega t} = g^{(1)} \cos \Omega t + g^{(2)} \sin \Omega t\,, \qquad (5.19)$$

often called frequency response, with period $T = 2\pi/\Omega$ and the amplitudes a_i and phase angles ψ_i:

$$a_i = \sqrt{g_i^{(1)2} + g_i^{(2)2}}\,, \quad \tan \psi_i = \frac{g_i^{(2)}}{g_i^{(1)}}\,, \quad i = 1(1)n\,. \qquad (5.20)$$

Obviously, the frequency response is completely characterized by the complex vector $g(\Omega)$.

Then, the response (5.10) to periodic forces reads, for the first harmonic, as

(5.21) $$\mathbf{x}_p(t) = \Phi(t)\left[\mathbf{x}_0 - \mathbf{x}_\infty - \mathbf{g}^{(1)}\right] + \mathbf{x}_\infty + \mathbf{x}_{p\infty}(t).$$

The steady-state behavior to periodic excitations depends again on the stability of the vibration system (5.3):

$\mathbf{x}_p(t \to \infty)$ is bounded if (5.3) asymptotically stable,

$\mathbf{x}_p(t \to \infty)$ may be bounded or unbounded if (5.3) stable,

$\mathbf{x}_p(t \to \infty)$ is unbounded if (5.3) unstable.

The steady-state response of an asymptotically stable system follows from (5.21) as

(5.22) $$\mathbf{x}_p(t \to \infty) = \mathbf{x}_\infty + \mathbf{x}_{p\infty}(t) .$$

Thus, the frequency response (5.19) is the most essential part of the steady-state response. Therefore, the frequency response, characterized by the vector $\mathbf{g}(\Omega)$ and the frequency matrix $\mathbf{F}(\Omega)$, will be investigated and intrepreted in detail in the following Chapter 6. Further, the components of the vector $\mathbf{g}$ can be represented graphically if the frequency Ω is used as an independent variable :

(5.23) $$g_i(\Omega) = a_i(\Omega)\, e^{-i\psi_i(\Omega)} , \quad i = 1(1)n$$

where $a_i(\Omega)$ is the amplitude function, $\psi_i(\Omega)$ is the phase function. In the complex plane, $g_i(\Omega)$ is called the locus function. Some typical plots of these functions are shown in Figs. 5.3 - 5.5.

General statements for the amplitude, phase

and/or locus function are possible in the limit cases $\Omega = 0$ and $\Omega \to \infty$ and for the forcing function $\mathbf{f} = \mathbf{f}^{(1)} \cos \Omega t$. Then it follows from (5.17) and (5.18)

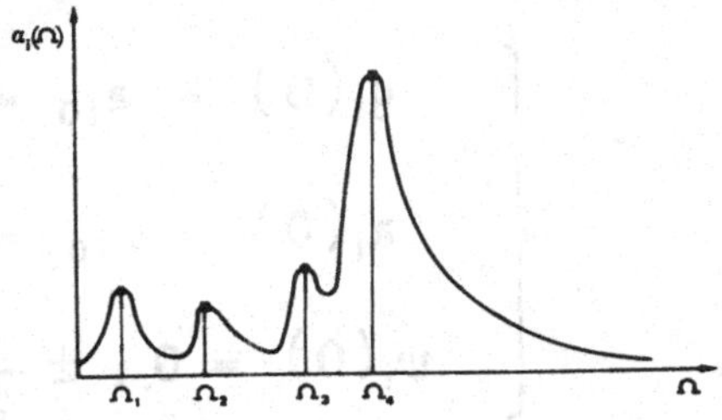

Fig.5.3. Typical amplitude plot

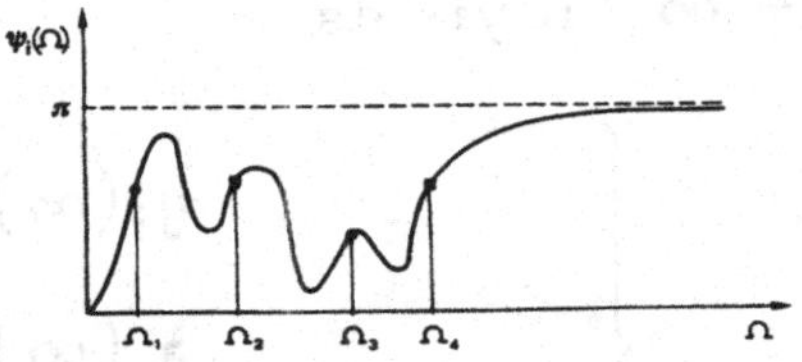

Fig.5.4. Typical phase plot

$$\mathbf{g} = (i\Omega \mathbf{E} - \mathbf{A})^{-1} \cdot \mathbf{f}^{(1)} =$$

$$= \frac{\operatorname{adj}(i\Omega \mathbf{E} - \mathbf{A})}{\det(i\Omega \mathbf{E} - \mathbf{A})} \mathbf{f}^{(1)} \tag{5.24}$$

or

$$g_i(\Omega) = \frac{b_{i1}(i\Omega)^{n-1} + b_{i2}(i\Omega)^{n-2} + \ldots + b_{in}}{(i\Omega)^n + a_1(i\Omega)^{n-1} + a_2(i\Omega)^{n-2} + \ldots + a_n} \,, \quad i = 1(1)n \,, \tag{5.25}$$

respectively, where a_k are the characteristic coefficients and b_{ik} are real coefficients depending on the matrix $\mathbf{A}$ and the real forcing vector $\mathbf{f}^{(1)}$, $k = 1(1)n$. For $\Omega = 0$, one obtains from (5.25)

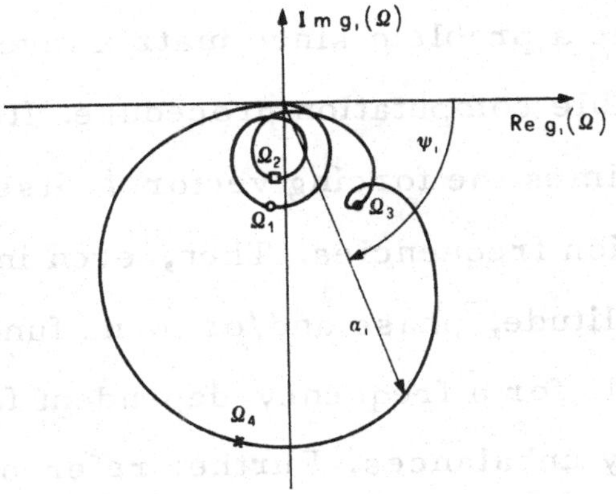

Fig.5.5. Typical polar plot of the locus function

$$(5.26)\quad \begin{cases} g_i(0) = a_{i0} = \dfrac{b_{in}}{a_n} = \text{const}\,, \\ a_i(0) = a_{i0} = \dfrac{b_{in}}{a_n} = \text{const}\,, \\ \psi_i(0) = 0\,, \pm\dfrac{\pi}{2}, \ldots, \pm\dfrac{\pi}{2}(n-1). \end{cases}$$

For $\Omega \to \infty$, it yields

$$(5.27)\quad \begin{cases} g_i(\infty) = 0\,, \\ a_i(\infty) = 0\,, \\ \psi_i(\infty) = \pm\dfrac{\pi}{2}\,, \pm\pi\,, \ldots, \pm\dfrac{\pi}{2}n\,. \end{cases}$$

Obviously, $\psi_i(\infty) = +\frac{\pi}{2}$ follows for $b_{i1} > 0$ while $\psi_i(\infty) = \frac{\pi}{2}n$ is obtained if $b_{i1} = b_{i2} = \ldots b_{in-1} = 0$ and $b_{in} > 0$. Beyond these general statements, the amplitude, phase and/or locus function has to be computed numerically. But this is not expected to be a problem since matrix inversion is an everywhere available computation procedure. It has to be mentioned that sometimes the forcing vector $\mathbf{f}$ itself is depending on the excitation frequencies. Then, even in the limit cases, various amplitude, phase and/or locus functions are obtained. An example for a frequency-dependent forcing vector is the excitation by unbalances. Further reference is given to Klotter (1951), (1960), Lippmann (1968) and Magnus, Lückel, Müller, Schiehlen (1971).

CHAPTER 6

Resonance and Absorption

Harmonic excitation often occurs in engineering systems. It is commonly produced by the unbalance in machines with rotating parts. Further, understanding the behavior of a system with harmonic excitation is an essential indication how the system will respond to more general types of excitation. Therefore, for system analysis the frequency response introduced in Chapter 5 has to be investigated in more detail. This investigation yields a theory of resonance and absorption affecting the design of vibration system little sensitive to excitations, the tuning of vibration absorbers, the contraction of vibration measuring instruments, etc.

6.1 Frequency Response

The frequency response of an asymptotically stable vibration system

$$\dot{\mathbf{x}}(t) = \mathbf{A}\mathbf{x}(t) + \mathbf{f}(t) \tag{6.1}$$

excited by a harmonic vector function (5.13)

(6.2) $$\mathbf{f}(t) = \mathbf{f}^{(1)}\cos\Omega t + \mathbf{f}^{(2)}\sin\Omega t$$

was given by (5.17-5.19)

(6.3) $$\mathbf{x}_{p\infty}(t) = \mathbf{g}^{(1)}\cos\Omega t + \mathbf{g}^{(2)}\sin\Omega t$$

where

(6.4) $$\mathbf{g}^{(1)} - i\mathbf{g}^{(2)} = \mathbf{F}(\Omega)\left(\mathbf{f}^{(1)} - i\mathbf{f}^{(2)}\right)$$

with the frequency response matrix

(6.5) $$\mathbf{F}(\Omega) = (i\Omega\mathbf{E} - \mathbf{A})^{-1}$$

The frequency response vectors $\mathbf{g}^{(1)}$ and $\mathbf{g}^{(2)}$ can also be calculated without complex manipulations

(6.6) $$\begin{aligned} \mathbf{g}^{(1)} &= -\left(\Omega^2\mathbf{E} + \mathbf{A}^2\right)^{-1}\left(\mathbf{A}\mathbf{f}^{(1)} + \Omega\mathbf{f}^{(2)}\right), \\ \mathbf{g}^{(2)} &= -\left(\Omega^2\mathbf{E} + \mathbf{A}^2\right)^{-1}\left(-\Omega\mathbf{f}^{(1)} + \mathbf{A}\mathbf{f}^{(2)}\right). \end{aligned}$$

If the mechanical vibration system is given by the second order equation

$$\mathbf{M}\ddot{\mathbf{y}}(t) + (\mathbf{D} + \mathbf{G})\dot{\mathbf{y}}(t) + (\mathbf{K} + \mathbf{N})\mathbf{y}(t) = \mathbf{h}^{(1)}\cos\Omega t + \mathbf{h}^{(2)}\sin\Omega t\,,$$
(6.7)

then, the steady-state response of the system is described by

(6.8) $$\mathbf{y}_{p\infty}(t) = \mathbf{q}^{(1)}\cos\Omega t + \mathbf{q}^{(2)}\sin\Omega t$$

where

$$q^{(1)} - iq^{(2)} = \left[-\Omega^2\mathbf{M} + i\Omega(\mathbf{D}+\mathbf{G}) + (\mathbf{K}+\mathbf{N})\right]^{-1}\left(\mathbf{h}^{(1)} - i\,\mathbf{h}^{(2)}\right) \qquad (6.9)$$

or equivalently

$$\begin{bmatrix} q^{(1)} \\ q^{(2)} \end{bmatrix} = \begin{bmatrix} -\Omega^2\mathbf{M}+(\mathbf{K}+\mathbf{N}) & \Omega(\mathbf{D}+\mathbf{G}) \\ -\Omega(\mathbf{D}+\mathbf{G}) & -\Omega^2\mathbf{M}+(\mathbf{K}+\mathbf{N}) \end{bmatrix}^{-1} \begin{bmatrix} \mathbf{h}^{(1)} \\ \mathbf{h}^{(2)} \end{bmatrix} . \qquad (6.10)$$

The response $\dot{\mathbf{y}}_{p\infty}(t)$ is easily obtained from (6.8) as

$$\dot{\mathbf{y}}_{p\infty}(t) = \Omega\,\mathbf{q}^{(2)}\cos\Omega t - \Omega\,\mathbf{q}^{(1)}\sin\Omega t . \qquad (6.11)$$

The representations (6.3 - 6.5) and (6.8 - 6.11) are equivalent for mechanical systems. For the further development the first order representation (6.3 - 6.5) will be used.
The steady-state response (6.3) is essentially determined by the frequency response matrix (6.5). Therefore, one has to discuss the properties of this matrix. Usually it is required that the amplitudes $a_i(\Omega)$ (5.20),

$$a_i = \sqrt{g_i^{(1)2} + g_i^{(2)2}} \quad , \quad i = 1(1)n \ , \qquad (6.12)$$

of the coordinates $x_{ip\infty}(t)$ will be small for some frequency domain of the harmonic ewcitation. As shown in Fig. 5.3 the amplitude functions $a_i(\Omega)$ have peaks at some critical frequencies. These peaks are determined by $\mathbf{F}(\Omega)$ and by the input vectors $\mathbf{f}^{(1)}$ and $\mathbf{f}^{(2)}$. Although asymptotic stability of the system matrix $\mathbf{A}$ implies boundedness of the steady-state response (theorem 4.3) one is interested in the maximal amplitudes of $a_i(\Omega)$, particularly when the damping of the vibration system is small or approaches zero. Therefore, in the following discussion stable matrices **A** will be included.

6.1.1 Elementary Frequency Responses

The general frequency response (6.3) can be interpreted as a superposition of elementary frequency responses. For the most important case of simple system matrices **A** the real nodal transformation (3.27) leads (6.1) to the nodal representation of the harmonically excited vibration system

$$\dot{\bar{\mathbf{x}}} = \bar{\bar{\Lambda}}\,\bar{\mathbf{x}} + \bar{\mathbf{X}}^{-1}\left(\mathbf{f}^{(1)}\cos\Omega t + \mathbf{f}^{(2)}\sin\Omega t\right) \tag{6.13}$$

where $\bar{\bar{\Lambda}}$ is a real block diagonal matrix of l real eigenvalues $\lambda_i = -\delta_i$, $i = 1(1)l$, and of $s = \frac{1}{2}(n-l)$ real 2×2 blocks

$$\begin{bmatrix} -\delta_k & \omega_k \\ -\omega_k & -\delta_k \end{bmatrix}$$

for the $n-l$ complex eigenvalues $\lambda_{k,k+1} = -\delta_k \pm i\omega_k$:

$$\bar{\Lambda} = \begin{bmatrix} -\delta_1 & & & & & & & 0 \\ & \ddots & & & & & & \\ & & -\delta_l & & & & & \\ & & & -\delta_{l+1} & \omega_{l+1} & & & \\ & & & -\omega_{l+1} & -\delta_{l+1} & & & \\ & & & & & \ddots & & \\ & & & & & & -\delta_{l+s} & \omega_{l+s} \\ 0 & & & & & & -\omega_{l+s} & -\delta_{l+s} \end{bmatrix} . \qquad (6.14)$$

The problem of the general frequency response is decoupled in $l + s$ problems of elementary frequency responses. These elementary frequency responses are represented by two different types. By (6.13) for real eigenvalues the scalar equation

$$\dot{\bar{\bar{x}}}_k = -\delta_k \bar{\bar{x}}_k + \bar{F}_k^{(1)} \cos \Omega t + \bar{F}_k^{(2)} \sin \Omega t \;, \qquad k = 1(1)l \;, \qquad (6.15)$$

is obtained while for a pair of complex eigenvalues the coupled equations

$$\begin{aligned} \dot{\bar{\bar{x}}}_k &= -\delta_k \bar{\bar{x}}_k + \omega_k \bar{\bar{x}}_{k+1} + \bar{F}_k^{(1)} \cos \Omega t + \bar{F}_k^{(2)} \sin \Omega t \;, \\ \dot{\bar{\bar{x}}}_{k+1} &= -\omega_k \bar{\bar{x}}_k - \delta_k \bar{\bar{x}}_{k+1} + \bar{F}_{k+1}^{(1)} \cos \Omega t + \bar{F}_{k+1}^{(2)} \sin \Omega t \;, \qquad (6.16) \\ & \qquad k = l+1\,(2)\,l+2s-1 \;, \end{aligned}$$

are relevant (compare 3.26a). For abbreviation $\bar{f}^{(1)} = \bar{X}^{-1} f^{(1)}$ and $\bar{f}^{(2)} = \bar{X}^{-1} f^{(2)}$ are introduced.

The frequency response of the elementary system (6.15) is given by

$$\bar{\bar{g}}_k^{(1)} - i\,\bar{\bar{g}}_k^{(2)} = \frac{\delta_k \bar{F}_k^{(1)} - \Omega \bar{F}_k^{(2)}}{\delta_k^2 + \Omega^2} - i\, \frac{\Omega \bar{F}_k^{(1)} + \delta_k \bar{F}_k^{(2)}}{\delta_k^2 + \Omega^2} \qquad (6.17)$$

yielding the amplitude function

$$\bar{\bar{a}}_k^2(\Omega) = \frac{\bar{F}_k^{(1)2} + \bar{F}_k^{(2)2}}{\delta_k^2 + \Omega^2} \tag{6.18}$$

and the phase function

$$\tan \bar{\bar{\psi}}_k(\Omega) = \frac{\Omega \bar{F}_k^{(1)} + \delta_k \bar{F}_k^{(2)}}{\delta_k \bar{F}_k^{(1)} - \Omega \bar{F}_k^{(2)}} \quad , \qquad k = 1(1)l. \tag{6.19}$$

In a similar way the frequency response of the second elementary system (6.16) is calculated

$$\bar{\bar{g}}_k^{(1)} - i\,\bar{\bar{g}}_k^{(2)} = \frac{1}{(\delta_k^2 + \omega_k^2 - \Omega^2)^2 + 4\delta_k^2\Omega^2} \,\cdot$$

$$\cdot \begin{bmatrix} \delta_k(\delta_k^2 + \omega_k^2 + \Omega^2) & \omega_k(\delta_k^2 + \omega_k^2 - \Omega^2) \\ -i\Omega(\delta_k^2 - \omega_k^2 + \Omega^2) & -2i\delta_k\omega_k\Omega \\ -\omega_k(\delta_k^2 + \omega_k^2 - \Omega^2) & \delta_k(\delta_k^2 + \omega_k^2 + \Omega^2) \\ +2i\delta_k\omega_k\Omega & -i\Omega(\delta_k^2 - \omega_k^2 + \Omega^2) \end{bmatrix} \cdot \begin{bmatrix} \bar{F}_k^{(1)} - i\bar{F}_k^{(2)} \\ \bar{F}_{k+1}^{(1)} - i\bar{F}_{k+1}^{(2)} \end{bmatrix}. \tag{6.20}$$

For example, the amplitude functions of the two coordinates are

$$\bar{\bar{a}}_k^2(\Omega) = \frac{(\delta_k \bar{F}_k^{(1)} + \omega_k \bar{F}_{k+1}^{(1)} + \Omega \bar{F}_k^{(2)})^2 + (\delta_k \bar{F}_k^{(2)} + \omega_k \bar{F}_{k+1}^{(2)} - \Omega \bar{F}_k^{(1)})^2}{(\delta_k^2 + \omega_k^2 - \Omega^2)^2 + 4\delta_k^2\Omega^2}, \tag{6.21a}$$

$$\bar{\bar{a}}^{2}_{k+1}(\Omega) = \frac{\left(-\omega_k \bar{F}^{(1)}_k + \delta_k \bar{F}^{(1)}_{k+1} + \Omega \bar{F}^{(2)}_{k+1}\right)^2 + \left(-\omega_k \bar{F}^{(2)}_k + \delta_k \bar{F}^{(2)}_{k+1} - \Omega \bar{F}^{(1)}_{k+1}\right)^2}{\left(\delta_k^2 + \omega_k^2 - \Omega^2\right)^2 + 4\delta_k^2 \Omega^2} , \tag{6.21b}$$

$$k = l+1(2)l+2s-1 .$$

For illustration of these results a one-degree-of-freedom mechanical system is considered :

$$m\ddot{y} + d\dot{y} + cy = e \cos \Omega t . \tag{6.22}$$

Using the abbreviations

$$2\delta = \frac{d}{m} , \quad \delta^2 + \omega^2 = \frac{c}{m}$$

the state space equation of (6.22) reads as

$$\dot{x}(t) = \begin{bmatrix} 0 & 1 \\ -(\delta^2+\omega^2) & -2\delta \end{bmatrix} x(t) + \begin{bmatrix} 0 \\ \frac{e}{m} \end{bmatrix} \cos \Omega t . \tag{6.23}$$

Assume conjugate complex eigenvalues, i. e. $\lambda_{1,2} = -\delta \pm i\omega$ and $\Delta = \frac{\delta}{\sqrt{\delta^2+\omega^2}} < 1$. Then the real modal transformation $x(t) = \bar{X}\,\bar{x}(t)$ where

$$\bar{X} = \begin{bmatrix} 1 & 0 \\ -\delta & \omega \end{bmatrix} , \quad \bar{X}^{-1} = \begin{bmatrix} 1 & 0 \\ \frac{\delta}{\omega} & \frac{1}{\omega} \end{bmatrix} , \tag{6.24}$$

results in

$$\dot{\bar{x}} = \begin{bmatrix} -\delta & \omega \\ -\omega & \delta \end{bmatrix} \bar{x}(t) + \begin{bmatrix} 0 \\ \frac{e}{m\omega} \end{bmatrix} \cos \Omega t . \tag{6.25}$$

This system (6.25) is of type (6.16). Therefore, the frequency response of (6.25) is directly given by (6.20) for $\bar{F}_1^{(1)} = \bar{F}^{(2)} = \bar{F}_2^{(2)} = 0$ and $\bar{F}_2^{(1)} = \frac{e}{m\omega}$. The back transformation in physical coordinates is obtained by

(6.26) $$g^{(1)} - i g^{(2)} = \bar{X}\left(\bar{\bar{g}}^{(1)} - i \bar{\bar{g}}^{(2)}\right).$$

Finally, one gets from (6.20) and (6.26)

(6.27)
$$g^{(1)} = \frac{e}{m} \frac{1}{(\delta^2+\omega^2-\Omega^2)^2 + 4\delta^2\Omega^2} \begin{bmatrix} \delta^2+\omega^2-\Omega^2 \\ 2\delta\Omega^2 \end{bmatrix},$$

$$g^{(2)} = \frac{e}{m} \frac{1}{(\delta^2+\omega^2-\Omega^2)^2 + 4\delta^2\Omega^2} \begin{bmatrix} 2\delta\Omega \\ -\Omega(\delta^2+\omega^2-\Omega^2) \end{bmatrix}.$$

This leads to the well-known amplitude and phase functions for the displacement and for the velocity coordinate, y and $\dot{y}$, respectively :

(6.28a) $$a_1(\Omega) = \frac{e}{m} \frac{1}{\sqrt{(\delta^2+\omega^2-\Omega^2)^2 + 4\delta^2\Omega^2}},$$

(6.28b) $$a_2(\Omega) = \frac{e}{m} \frac{\Omega}{\sqrt{(\delta^2+\omega^2-\Omega^2)^2 + 4\delta^2\Omega^2}} = \Omega\, a_1(\Omega),$$

(6.29a) $$\tan\psi_1(\Omega) = \frac{2\delta\Omega}{\delta^2+\omega^2-\Omega^2},$$

$$\tan\psi_2(\Omega) = -\frac{\delta^2+\omega^2-\Omega^2}{2\delta\Omega} = \tan\left(\psi_1 - \frac{\pi}{2}\right). \qquad (6.29b)$$

The amplitude functions characterize the magnitudes of the steady-state vibration response of $y_{p\infty}(t)$ and $\dot{y}_{p\infty}(t)$ dependent on the excitation frequency Ω, while the phase functions give the shifting of the phase angles of the displacement and the velocity ~~with respect to the~~ input. The qualitative curves of the functions (6.28a - 6.29b) are plotted in Figs. 6.1 - 6.6 for a constant excitation amplitude e and various values of the damping ratio $\Delta = \frac{\delta}{\sqrt{\delta^2+\omega^2}} = \frac{d}{2\sqrt{cm}}$.

In Figs. 6.1 and 6.4 there are dotted lines representing the geometrical locus of maximal values of the amplitude functions for various damping ratios Δ:

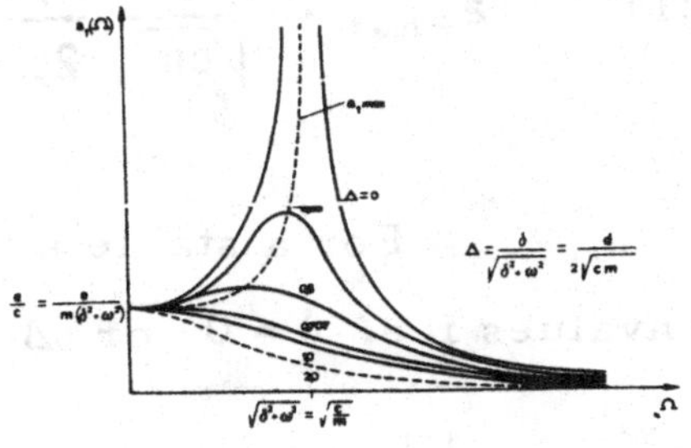

Fig.6.1. Displacement amplitude plot for various values of damping ratio

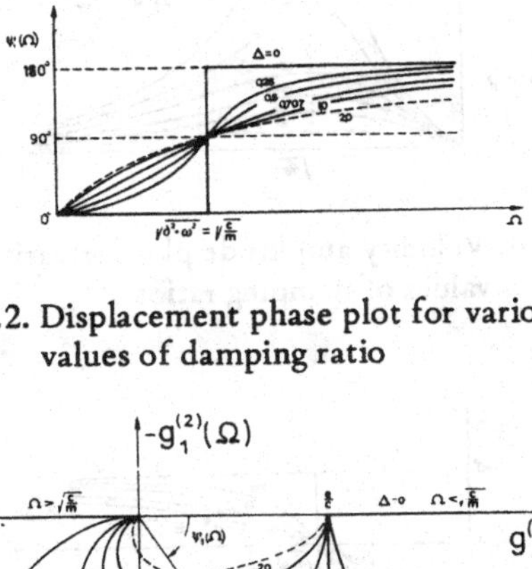

Fig.6.2. Displacement phase plot for various values of damping ratio

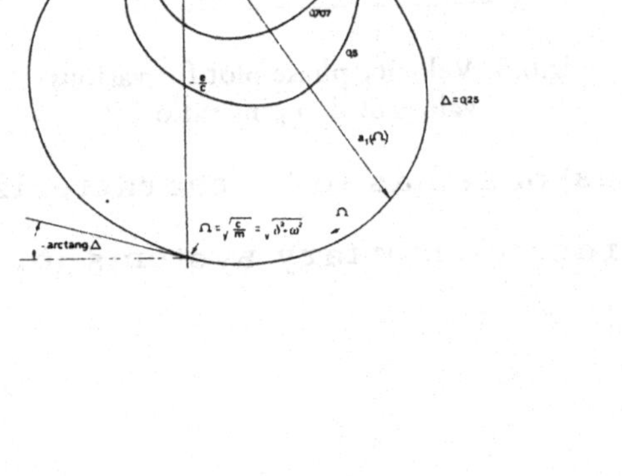

Fig.6.3. Displacement polar plot for various values of damping ratio

$$
(6.30)\qquad a_{1\,max} = \begin{cases} \dfrac{e}{c}\,\dfrac{1}{\sqrt{1-\left(\dfrac{\Omega^2}{\delta^2+\omega^2}\right)^2}} = \dfrac{e}{c}\,\dfrac{1}{2\Delta}\,\dfrac{1}{\sqrt{1-\Delta^2}} \\ \qquad\qquad \text{for } \Omega^2 = \dfrac{\delta^2+\omega^2}{1-2\Delta^2}\,,\ \Delta < \dfrac{1}{\sqrt{2}}\,, \\ \dfrac{e}{c} \text{ for } \Omega = 0\,,\ \Delta \geq \dfrac{1}{\sqrt{2}}\,, \end{cases}
$$

$$
(6.31)\qquad a_{2\,max} = \frac{e}{\sqrt{cm}}\,\frac{1}{2\Delta} \quad \text{for} \quad \Omega = \sqrt{\frac{c}{m}} = \sqrt{\delta^2+\omega^2}\,.
$$

For a stable system with purely imaginary eigen-eigenvalues i.e. $\delta = 0$ or $\Delta = 0$, the maximum amplitude approaches infinity.

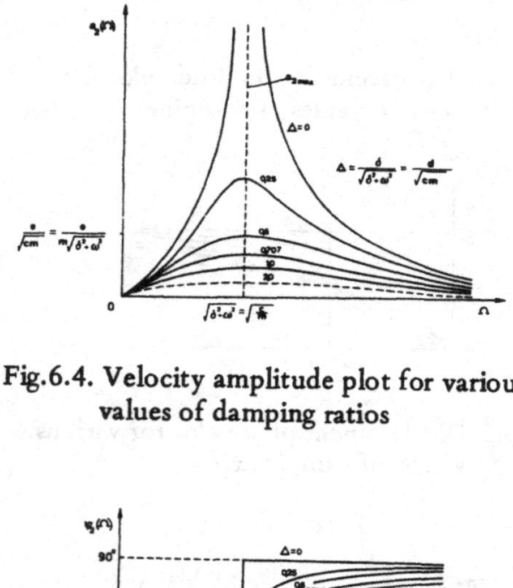

Fig.6.4. Velocity amplitude plot for various values of damping ratios

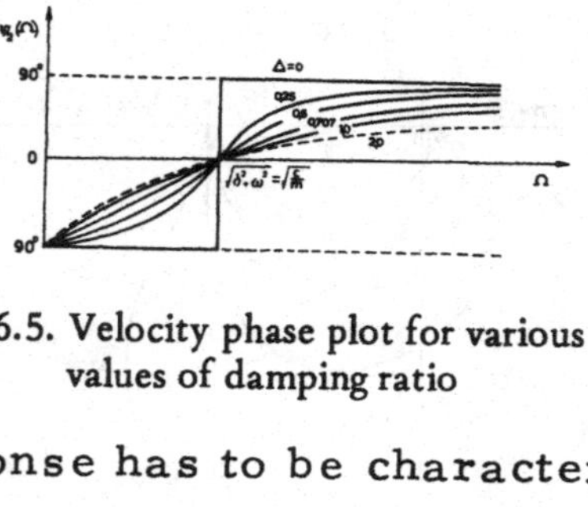

Fig.6.5. Velocity phase plot for various values of damping ratio

The curve marked by little lines characterizes an amplitude function for $\Delta = 2$, i.e. for a mechanical system (6.22) with two real eigenvalues. Although the results (6.27 - 6.29) were derived from the second elementary system (6.16) with complex eigenvalues, the frequency response (6.27) remains valid. But in general systems with real eigenvalues the frequency response has to be characterized by the superposition of first order elementary systems (6.15). For illustration amplitude, phase and locus functions are shown for $\bar{F}_K^{(1)} = 1$ and $\bar{F}_K^{(2)} = 0$ in Figs. 6.7 - 6.9.

The amplitude, phase and locus functions shown in Figs. 6.1 - 6.9 are characteristic for excitations with input vectors $\mathbf{f}^{(1)}$ and $\mathbf{f}^{(2)}$ independent on Ω. If, for example, dynamical mass unbalances lead to input vectors increasing with Ω^2, then quite different amplitude, phase and polar plots are obtained.

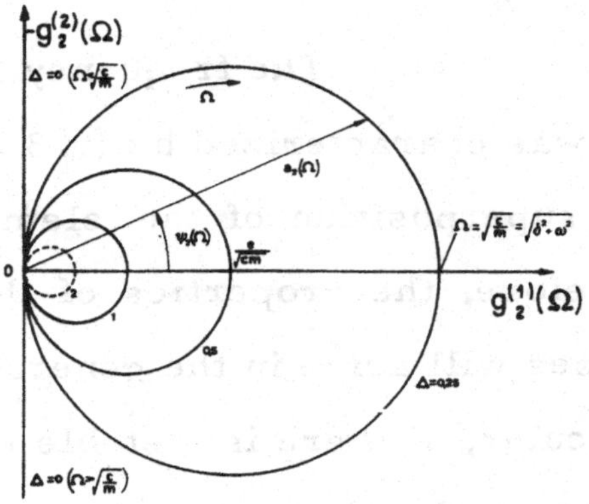

Fig.6.6. Velocity polar plot for various values of damping ratio

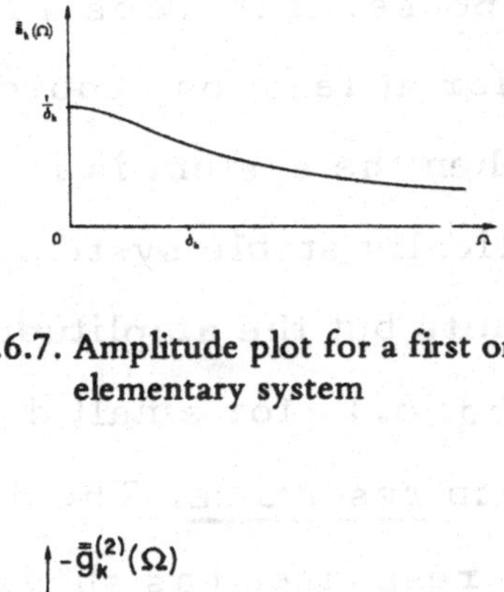

Fig.6.7. Amplitude plot for a first order elementary system

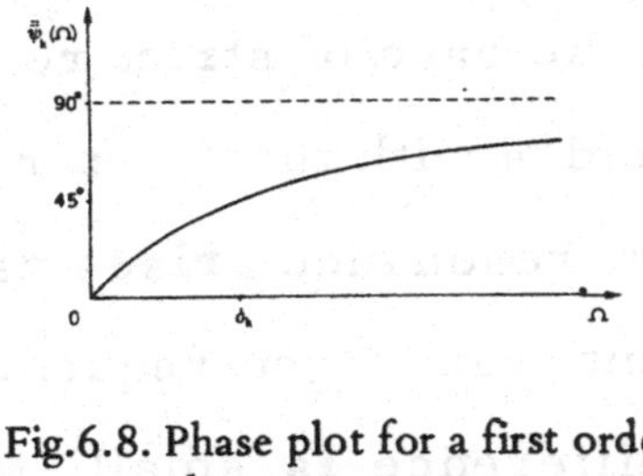

Fig.6.8. Phase plot for a first order elementary system

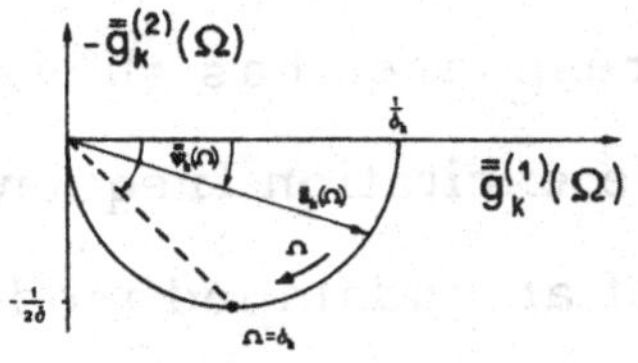

Fig.6.9. Polar plot for a first order elementary system

6.1.2 General Frequency Response

The frequency response of the system (6.1, 6.2) was characterized by (6.3 - 6.5) and is obtained by a suitable superposition of the elementary frequency responses. Therefore, the properties of the elementary frequency responses will arise in the general frequency response, too. In particular, if there is a stable (but not asymptotically stable) mode then the elementary frequency response approaches infinity for an excitation frequency equal to the eigenfrequency and this may cause a similar behavior of the general frequency response. If there is a frequency Ω implying an infinite value for at least one coordinate of the steady-state response (6.3) then the system is in <u>strict resonance</u>. For a weakly asymptotically stable system, however, the amplitude will not be infinite but the amplitude function peaks will be large, see e.g. Fig. 6.1 for small damping ratios Δ. Then the system is in <u>resonance</u>. The discussion of the elementary frequency responses has shown that in the case of strict resonance the excitation frequency coincides with the eigenfrequency of an undamped mode but that resonance arises usually for excitation frequencies different from eigenfrequencies. However, for small damping this difference is small, too. Since the investigation of resonances is very cumbersome, and since resonance phenomena are continuous functions of the parameters of the dynamical system, usually strict resonance is investigated instead of resonance for small damping.

Due to the linear superposition of the elementary frequency responses also new effects exist for the general frequency response: there may occur some cancelling phenomena. The amplitude functions (6.18, 6.28a, b) do not vanish for $0 < \Omega < \infty$, but in the general case elementary frequency responses may cancel themselves for a certain Ω . Cancellation cannot occur in a first order system (6.15) or in a one-degree-of-freedom mechanical system (6.22) but it can still appear in a general second order system (6.16): Choosing $\bar{F}_k^{(1)} = \omega_k^2$, $\bar{F}_k^{(2)} = -\omega_k \delta_k$, $F_{k+1}^{(1)} = 0$ and $F_{k+1}^{(2)} = \delta_k^2 + \omega_k^2$ the amplitude function $\bar{a}_k = (\Omega)$ (6.21) vanishes for $\Omega = \omega_k$ and $\delta_k > 0$. This vanishing effect is called an absorption. Another phenomenon may happen by a simultaneous occurrence of strict resonance and of absorption: choosing additionally $\delta_k = 0$ in the example of $\bar{\bar{a}}_k(\Omega)$ numerator and denominator vanish for $\Omega = \omega_k$ and yield a finite limit: $\bar{\bar{a}}_k(\Omega \to \omega_k) = \frac{1}{2}\omega_k$. The strict resonance is cancelled by an absorption. Therefore, this effect is called pseudo-resonance.

From the mathematical point of view the effects of strict resonance, absorption, and pseudo-resonance are a problem of a vanishing denominator, a vanishing numerator, or of both within the frequency response. Therefore, the solution (6.4) for the frequency response $\mathbf{g}^{(1)}$ and $\mathbf{g}^{(2)}$ has

has to be presented in the following form

$$(6.32) \qquad \mathbf{g}^{(1)} - i\mathbf{g}^{(2)} = \frac{\operatorname{adj}(i\Omega\mathbf{E}-\mathbf{A})(\mathbf{f}^{(1)} - i\mathbf{f}^{(2)})}{\det(i\Omega\mathbf{E}-\mathbf{A})}$$

where

$$(6.33) \qquad \mathbf{F}(\Omega) = (i\Omega\mathbf{E}-\mathbf{A})^{-1} = \frac{\operatorname{adj}(i\Omega\mathbf{E}-\mathbf{A})}{\det(i\Omega\mathbf{E}-\mathbf{A})}$$

has been used. Here, $\operatorname{adj}(i\Omega\mathbf{E}-\mathbf{A})$ is the so-called adjoint matrix of $(i\Omega\mathbf{E}-\mathbf{A})$. In (6.32) the frequency response is written as a numerator vector divided by a scalar denominator polynomial. By this representation the effects of resonances and absorptions will be discussed in the next section in more detail.

6.2 Strict Resonance, Pseudo-Resonance, and Absorption Criteria

First, some simple considerations for the criteria of strict resonance, of pseudo-resonance, and of absorption are given. A necessary condition for strict resonance is the vanishing of the characteristic polynomial (3.16) for $\lambda = i\Omega_R$

$$(6.34) \qquad p(i\Omega_R) = \det(i\Omega_R\mathbf{E}-\mathbf{A}) = 0$$

where Ω_R is the resonance frequency. Therefore, strict resonance does not occur in asymptotically stable systems. For sufficient conditions the numerator vector $\mathrm{adj}(i\Omega_R\mathbf{E}-\mathbf{A})(\mathbf{f}^{(1)}-i\mathbf{f}^{(2)})$ has to be considered, too. Also for vibration absorption and for pseudo-resonance this numerator is relevant.

The numerator of (6.32) consists essentially of the adjoint matrix, $\mathrm{adj}(i\Omega\mathbf{E}-\mathbf{A})$. The adjoint matrix is defined as the transposed matrix of the cofactors of $(i\Omega\mathbf{E}-\mathbf{A})$, see Lancaster (1969). The following properties are summarized:

1.
$$(i\Omega\mathbf{E}-\mathbf{A})\,\mathrm{adj}\,(i\Omega\mathbf{E}-\mathbf{A}) = \mathrm{adj}\,(i\Omega\mathbf{E}-\mathbf{A})(i\Omega\mathbf{E}-\mathbf{A}) = \det(i\Omega\mathbf{E}-\mathbf{A})\cdot\mathbf{E} \tag{6.35}$$

2. (6.36)

rank $(i\Omega\mathbf{E}-\mathbf{A})$	rank adj$(i\Omega\mathbf{E}-\mathbf{A})$
n	n
$n-1$	1
$< n-1$	0

3. $\mathrm{rank}\left[\mathrm{adj}(i\Omega\mathbf{E}-\mathbf{A})\right]=1$:

a) $(i\Omega\mathbf{E}-\mathbf{A})\mathbf{x}=0$ has 1 independent solution vector presented by each nonvanishing colum vector of

$$\mathrm{adj}(i\Omega\mathbf{E}-\mathbf{A}), \tag{6.37a}$$

b) $\mathrm{adj}(i\Omega\mathbf{E}-\mathbf{A})\mathbf{x}=0$ has $n-1$ independent solution vectors presented by the $n-1$ independent column vectors of $(i\Omega\mathbf{E}-\mathbf{A})$. (6.37b)

Depending on $\mathbf{adj}(i\Omega_R\mathbf{E}-\mathbf{A})(\mathbf{f}^{(1)}-i\mathbf{f}^{(2)})$ the resonance frequency Ω_R yields strict or pseudo-resonance. Rank $(i\Omega_R\mathbf{E}-\mathbf{A}) = n-1$ and $\mathbf{f}^{(1)}-i\mathbf{f}^{(2)} \neq (i\Omega_R\mathbf{E}-\mathbf{A})\mathbf{k}$, $\mathbf{k}$ arbitrary, imply at least in one coordinate strict resonance. Further, rank $(i\Omega_R\mathbf{E}-\mathbf{A}) = n-1$ and $\mathbf{f}^{(1)}-i\mathbf{f}^{(2)} = (i\Omega_R\mathbf{E}-\mathbf{A})\mathbf{k}$ yield pseudo-resonance. It remains to discuss the case rank $(i\Omega_R\mathbf{E}-\mathbf{A}) < n-1$ which will be done later.

Absorption is guaranteed if $\det(i\Omega_A\mathbf{E}-\mathbf{A}) \neq 0$ and at least one of the coordinates of $\mathbf{adj}(i\Omega_A\mathbf{E}-\mathbf{A})(\mathbf{f}^{(1)}-i\mathbf{f}^{(2)})$ is zero. It should be noted that vibration absorption in all coordinates is not possible because $\det(i\Omega_A\mathbf{E}-\mathbf{A}) \neq 0$ implies a regular adjoint matrix and $\mathbf{g}^{(1)}-i\mathbf{g}^{(2)}$ vanishes if and only if $\mathbf{f}^{(1)}-i\mathbf{f}^{(2)} = \mathbf{0}$. However, it is remarkable that absorption arises in asymptotically stable systems. The discussion of absorber effects is more complicated if $\det(i\Omega_A\mathbf{E}-\mathbf{A}) = 0$, i.e. $\Omega_A = \Omega_R$. Then, absorption is only possible if there is a pseudo-resonance with vanishing amplitudes.

Investigating the critical case rank $(i\Omega_R\mathbf{E}-\mathbf{A}) < n-1$ the concept of generalized inverse is advantageous. The generalized inverse $\mathbf{B}^+$ of an arbitrary (rectangular) matrix $\mathbf{B}$ is uniquely defined as the solution of the four simultaneous equations (see Lancaster (1969))

$$\mathbf{B}^+\mathbf{B} = (\mathbf{B}^+\mathbf{B})^T, \qquad (6.38a)$$
$$\mathbf{B}\mathbf{B}^+ = (\mathbf{B}\mathbf{B}^+)^T,$$

$$\mathbf{B}\,\mathbf{B}^+\mathbf{B} = \mathbf{B},$$
$$\mathbf{B}^+\mathbf{B}\,\mathbf{B}^+ = \mathbf{B}^+. \tag{6.38b}$$

For an $n \times n$ regular matrix $\mathbf{B}$ the pseudo-inverse $\mathbf{B}^+$ coincides with the inverse $\mathbf{B}^{-1}$. An application of the generalized inverses is the solution of linear equations:

$$\mathbf{B}\,\mathbf{x} = \mathbf{b} \tag{6.39}$$

is solvable if and only if

$$\left(\mathbf{E} - \mathbf{B}\,\mathbf{B}^+\right)\mathbf{b} = \mathbf{0} \tag{6.40a}$$

or equivalently $\mathbf{b}$ is of the type

$$\mathbf{b} = \mathbf{B}\,\mathbf{k}\ ; \tag{6.40b}$$

then the solution is

$$\mathbf{x} = \mathbf{B}^+\mathbf{b} + \left(\mathbf{E} - \mathbf{B}^+\mathbf{B}\right)\tilde{\mathbf{x}} \tag{6.41a}$$

with an arbitrary vector $\tilde{\mathbf{x}}$; especially the minimum norm solution is

$$\mathbf{x} = \mathbf{B}^+\mathbf{b}. \tag{6.41b}$$

The problem of critical resonance frequency Ω_R is completely solved by this concept of generalized inverses. The frequency response vectors $\mathbf{g}^{(1)}$ and $\mathbf{g}^{(2)}$ satisfy

$$(6.42)\qquad (i\Omega\mathbf{E}-\mathbf{A})\left(\mathbf{g}^{(1)}-i\mathbf{g}^{(2)}\right)=\left(\mathbf{f}^{(1)}-i\mathbf{f}^{(2)}\right).$$

There is a (finite) solution of (6.42) if and only if

$$(6.43)\qquad \left[\mathbf{E}-(i\Omega\mathbf{E}-\mathbf{A})(i\Omega\mathbf{E}-\mathbf{A})^{+}\right]\left(\mathbf{f}^{(1)}-i\mathbf{f}^{(2)}\right)=0.$$

For Ω with $\det(i\Omega\mathbf{E}-\mathbf{A})\neq 0$ relation (6.43) is satisfied for each $\mathbf{f}^{(1)}-i\mathbf{f}^{(2)}$. No resonance or pseudo-resonance occur. For $\Omega=\Omega_R$ with $\det(i\Omega_R\mathbf{E}-\mathbf{A})=0$ there is no resonance if and only if (6.43) is valid, or equivalently

$$(6.44)\qquad \mathbf{f}^{(1)}-i\mathbf{f}^{(2)}=(i\Omega_R\mathbf{E}-\mathbf{A})\mathbf{k},$$

$\mathbf{k}$ arbitrary. Then the finite resonance response is characterized by the minimum norm solution (6.41 b)

$$(6.45)\qquad \begin{aligned}\mathbf{g}^{(1)}(\Omega_R)-i\mathbf{g}^{(2)}(\Omega_R)&=(i\Omega_R\mathbf{E}-\mathbf{A})^{+}\left(\mathbf{f}^{(1)}-i\mathbf{f}^{(2)}\right)=\\&=(i\Omega_R\mathbf{E}-\mathbf{A})^{+}(i\Omega_R\mathbf{E}-\mathbf{A})\mathbf{k}=\tilde{\mathbf{k}}.\end{aligned}$$

Table 6.1. Conditions for strict resonance, pseudo-resonance and absorbtion

		$\left[\mathbf{E}-(i\Omega\mathbf{E}-\mathbf{A})(i\Omega\mathbf{E}-\mathbf{A})^{+}\right](\mathbf{f}^{(1)}-i\mathbf{f}^{(2)})$ $= 0$: $\mathbf{f}^{(1)}-i\mathbf{f}^{(2)}=(i\Omega\mathbf{E}-\mathbf{A})\mathbf{k}$, $\tilde{\mathbf{k}}=(i\Omega\mathbf{E}-\mathbf{A})^{+}(i\Omega\mathbf{E}-\mathbf{A})\mathbf{k}$	$\left[\mathbf{E}-(i\Omega\mathbf{E}-\mathbf{A})(i\Omega\mathbf{E}-\mathbf{A})^{+}\right](\mathbf{f}^{(1)}-i\mathbf{f}^{(2)})$ $\neq 0$: $\mathbf{f}^{(1)}-i\mathbf{f}^{(2)}\neq(i\Omega\mathbf{E}-\mathbf{A})\mathbf{k}$
$\det(i\Omega\mathbf{E}-\mathbf{A})$	$=0$	$\tilde{k}_i\neq 0$: Pseudo-resonance $\tilde{k}_i=0$: Absorption	Strict resonance (at least in one coordinate)
$\det(i\Omega\mathbf{E}-\mathbf{A})$	$\neq 0$	$\tilde{\mathbf{k}}=\mathbf{k}=\mathbf{g}^{(1)}-i\mathbf{g}^{(2)}=\dfrac{\operatorname{adj}(i\Omega\mathbf{E}-\mathbf{A})(\mathbf{f}^{(1)}-i\mathbf{f}^{(2)})}{\det(i\Omega\mathbf{E}-\mathbf{A})}$ $k_i\neq 0$ Vibration $k_i=0$ Absorption	

Pseudo-resonance in the whole system is obtained if and only if (6.44) is true. Furthermore, vibration absorption occurs if some coordinates of $\tilde{k}$ are zero.

The results of this section can be summarized in Table 6.1.

6.3 Examples

Example 6.1: Automobile wheel suspension

Consider the automobile wheel suspension of example 2.1 where the vehicle is driving on a rough road, see Fig. 6.10. The mathematical model was represented by the second order equation

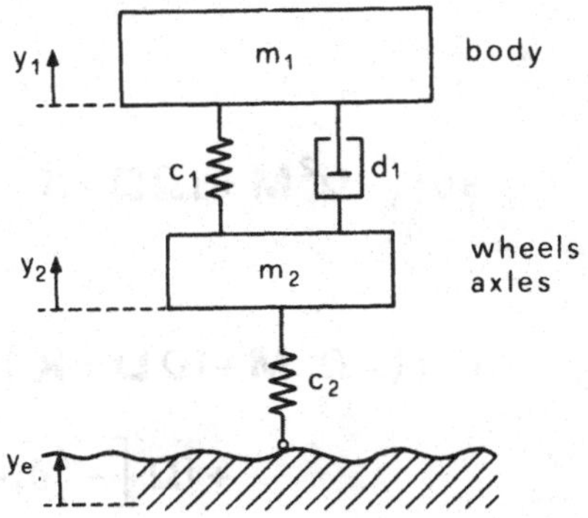

Fig.6.10. Automobile wheel suspension

$$\underbrace{\begin{bmatrix} m_1 & 0 \\ 0 & m_2 \end{bmatrix}}_{\mathbf{M}} \begin{bmatrix} \ddot{y}_1 \\ \ddot{y}_2 \end{bmatrix} + \underbrace{\begin{bmatrix} d_1 & -d_1 \\ -d_1 & d_1 \end{bmatrix}}_{\mathbf{D}} \begin{bmatrix} \dot{y}_1 \\ \dot{y}_2 \end{bmatrix} + \underbrace{\begin{bmatrix} c_1 & -c_1 \\ -c_1 & c_1+c_2 \end{bmatrix}}_{\mathbf{K}} \begin{bmatrix} y_1 \\ y_2 \end{bmatrix} = \underbrace{\begin{bmatrix} 0 \\ c_2 \end{bmatrix}}_{\mathbf{h}} y_e \, .$$

Assuming a harmonic excitation

$$y_e(t) = y_{e0} \cos \Omega t$$

the frequency response of the wheel suspension is easily obtained by (6.9). Introducing the abbreviations

$$\omega_1^2 = \frac{c_1}{m_1}\ ,\quad \omega_{12}^2 = \frac{c_1}{m_2}\ ,\quad \omega_2^2 = \frac{c_1+c_2}{m_2}\ ,\quad \delta_1 = \frac{d_1}{m_1}\ ,\quad \delta_2 = \frac{d_1}{m_2}$$

the frequency response matrix of the mechanical system is calculated as

$$(-\Omega^2\mathbf{M} + i\Omega\mathbf{D} + \mathbf{K})^{-1} = \frac{\operatorname{adj}(-\Omega^2\mathbf{M}+i\Omega\mathbf{D}+\mathbf{K})}{\det(-\Omega^2\mathbf{M}+i\Omega\mathbf{D}+\mathbf{K})}\ ,$$

$$\operatorname{adj}(-\Omega^2\mathbf{M}+i\Omega\mathbf{D}+\mathbf{K}) = \begin{bmatrix} \omega_2^2-\Omega^2+i\delta_2\Omega & \omega_1^2+i\delta_1\Omega \\ \omega_{12}^2+i\delta_2\Omega & \omega_1^2-\Omega^2+i\delta_1\Omega \end{bmatrix}\begin{bmatrix} m_2 & 0 \\ 0 & m_1 \end{bmatrix}$$

$$\det(-\Omega^2\mathbf{M}+i\Omega\mathbf{D}+\mathbf{K}) = m_1 m_2\Big\{(\omega_1^2-\Omega^2)(\omega_2^2-\Omega^2)-\omega_1^2\omega_{12}^2 + \\ + i\Omega\left[-(\delta_1+\delta_2)\Omega^2+\delta_1(\omega_2^2-\omega_{12}^2)\right]\Big\}\ .$$

This leads to the complex frequency response vector (6.9)

$$\mathbf{q}^{(1)} - i\mathbf{q}^{(2)} =$$

$$= \frac{y_{eo}(\omega_2^2-\omega_{12}^2)}{(\omega_1^2-\Omega^2)(\omega_2^2-\Omega^2)-\omega_1^2\omega_{12}^2+i\Omega\left[-(\delta_1+\delta_2)\Omega^2+\delta_1(\omega_2^2-\omega_{12}^2)\right]}\begin{bmatrix} \omega_1^2+i\delta_1\Omega \\ \omega_1^2-\Omega^2+i\delta_1\Omega \end{bmatrix}$$

which yields the amplitude functions

$$a_1^2(\Omega) = y_{eo}^2\,\frac{(\omega_2^2-\omega_{12}^2)^2(\omega_1^4+\delta_1^2\Omega^2)}{\left[(\omega_1^2-\Omega^2)(\omega_2^2-\Omega^2)-\omega_1^2\omega_{12}^2\right]^2+\Omega^2\left[-(\delta_1+\delta_2)\Omega^2+\delta_1(\omega_2^2-\omega_{12}^2)\right]^2}\ ,$$

$$a_2^2(\Omega) = y_{eo}^2 \frac{(\omega_2^2-\omega_{12}^2)^2\left[(\omega_1^2-\Omega^2)^2+\Omega^2\delta_1^2\right]}{\left[(\omega_1^2-\Omega^2)(\omega_2^2-\Omega^2)-\omega_1^2\omega_{12}^2\right]^2+\Omega^2\left[-(\delta_1+\delta_2)\Omega^2+\delta_1(\omega_2^2-\omega_{12}^2)\right]^2} .$$

Now the phenomena of resonance, pseudo-resonance and absorption will be discussed. Resonance is possible if the damping vanishes, $\delta_1 = \delta_2 = 0$. Then the necessary condition (6.34) of resonance leads to $\Omega_R = \Omega_{R_{1,2}}$ where $\Omega_{R_{1,2}}$ are the zeros of

$$(\omega_1^2-\Omega_R^2)(\omega_2^2-\Omega_R^2)-\omega_1^2\omega_{12}^2 = 0 .$$

Since $\Omega_{R_{1,2}}^2 \neq \omega_1^2$, the numerators of $a_1^2(\Omega)$ and $a_2^2(\Omega)$ do not vanish for $\Omega = \Omega_{R_{1,2}}$. Therefore, for $\Omega_{R_{1,2}}$ the system is in strict resonance $(\delta_1 = \delta_2 = 0)$. Pseudo-resonance does not exist here, but absorption can arise in the second coordinate if $\delta_1 = 0$. For $\Omega^2 = \Omega_A^2 = \omega_1^2$ and $\delta_1 = 0$ the amplitude of the axle coordinate vanishes. The mass m_1 acts like a vibration absorber for the mass m_2 if the system is excited with the frequency of this absorber.

Another phenomenon in resonance theory is the effect of <u>fixed points</u> in the amplitude plot of $a_1(\Omega)$. For

$$\Omega_{F_2}^2 = \omega_2^2-\omega_{12}^2 \quad \text{and} \quad \Omega_{F_1,F_3}^2 = \frac{1}{2}\left[\left(2+\frac{m_1}{m_2}\right)\omega_1^2+\omega_2^2\right] \pm$$

$$\pm\sqrt{\frac{1}{4}\left[\left(2+\frac{m_1}{m_2}\right)\omega_1^2+\omega_2^2\right]^2-2\,\omega_1^2(\omega_2^2-\omega_{12}^2)}$$

the amplitudes are independent of the damping parameter d_1.

The damping has no influence to the motion of the body if the automobile is forced by a harmonic road excitation with frequencies $\Omega_{Fi}, i = 1\ 2\ 3$. This effect is very interesting and leads to some design techniques of vibratory systems, see section 6.4.

From the technical point of view the frequency response of the vertical vehicle acceleration $\ddot{y}_1$ is important. This response is relevant to the vehicle comfort. Therefore in Fig. 6.11 some typical amplitude functions $\Omega^2 a_1(\Omega)$ of the automobile acceleration are plotted.

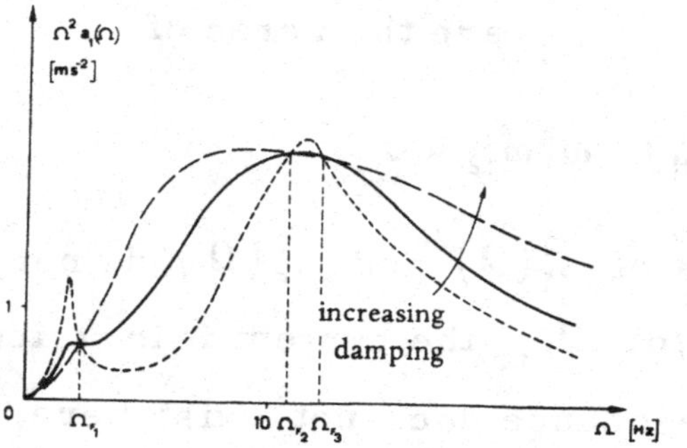

Fig.6.11. Amplitude plot of vertical automobile acceleration for various values of damping

The considered automobile wheel suspension results in a limited driving comfort due to the fixed points. Looking for better comfort the engine of the car can be designed as a vibration absorber. For this analysis the automobile is modeled by a body with m_2 and a linear suspension with spring coefficient c_2 and dashpot coefficient d_2 while the engine with mass m_1 is elastically coupled by a spring (coefficient c_1) to m_2. The masses of wheels and

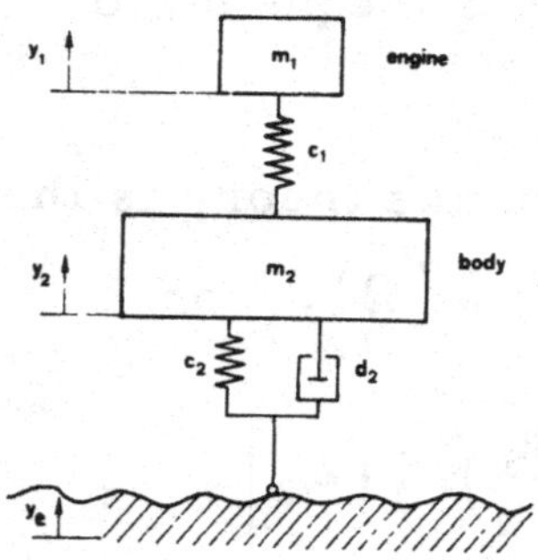

Fig.6.12 Engine absorber

axles as well as the tire are neglected. The model of such an engine absorber is sketched in Fig. 6.12.

The equation of motion reads as

$$\begin{bmatrix} m_1 & 0 \\ 0 & m_2 \end{bmatrix} \begin{bmatrix} \ddot{y}_1 \\ \ddot{y}_2 \end{bmatrix} + \begin{bmatrix} 0 & 0 \\ 0 & d_2 \end{bmatrix} \begin{bmatrix} \dot{y}_1 \\ \dot{y}_2 \end{bmatrix} + \begin{bmatrix} c_1 & -c_1 \\ -c_1 & c_1+c_2 \end{bmatrix} \begin{bmatrix} y_1 \\ y_2 \end{bmatrix} = \begin{bmatrix} 0 \\ 1 \end{bmatrix} (c_2 y_e + d_2 \dot{y}_e).$$

Assuming again a harmonic excitation

$$y_e(t) = y_{eo} \cos \Omega t$$

the amplitude functions of the steady-state frequency response are calculated as

$$a_1^2(\Omega) = y_{eo}^2 \frac{\omega_1^4 \left[(\omega_1^2 - \omega_{12}^2)^2 + \delta_2^2 \Omega^2\right]}{\left[(\omega_1^2 - \Omega^2)(\omega_2^2 - \Omega^2) - \omega_1^2 \omega_{12}^2\right]^2 + \delta_2^2 \Omega^2 (\omega_1^2 - \Omega^2)^2}$$

$$a_2^2(\Omega) = y_{eo}^2 \frac{(\omega_1^2 - \Omega^2)^2 \left[(\omega_1^2 - \omega_{12}^2)^2 + \delta_2^2 \Omega^2\right]}{\left[(\omega_1^2 - \Omega^2)(\omega_2^2 - \Omega^2) - \omega_1^2 \omega_{12}^2\right]^2 + \delta_2^2 \Omega^2 (\omega_1^2 - \Omega^2)^2}$$

where the abbreviations are

$$\omega_1^2 = \frac{c_1}{m_1}, \quad \omega_{12}^2 = \frac{c_1}{m_2}, \quad \omega_2^2 = \frac{c_1 + c_2}{m_2}, \quad \delta_2 = \frac{d_2}{m_2}.$$

The engine absorber is in strict resonance only for vanishing damping and the excitation frequencies $\Omega = \Omega_{R_{1,2}}$. The absorption effect occurs in the coordinate y_2 of the body;

$$\Omega_A^2 = \omega_1^2 \quad : \quad a_2^2(\omega_1) = 0 \; .$$

In contrary to the automobile wheel suspension absorption arises independently from the dashpot coefficient d_2. Therefore the absorption frequency Ω_A yields also a fixed point in the amplitude plot $a_2(\Omega)$. A qualitative sketch of $a_2(\Omega)$ is given in Fig. 6.13. Obviously, there exist three further fixed points. For

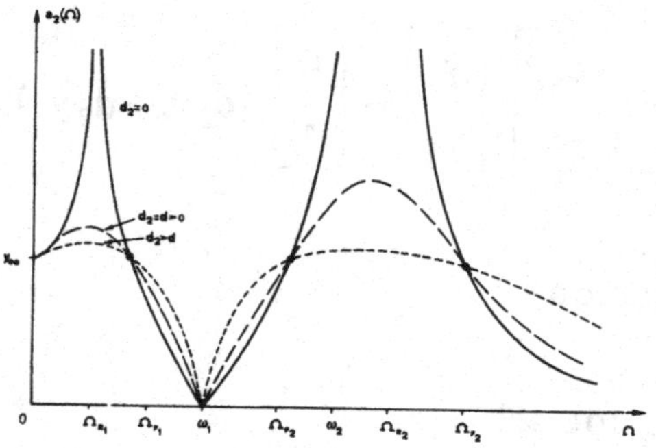

Fig.6.13. Amplitude plot of the automobile body frequency response using an engine absorber

$$\Omega_{F_2}^2 = \omega_1^2 + \omega_{12}^2 \quad \text{and} \quad \Omega_{F_1,F_3}^2 = \frac{1}{2}\left(2\omega_2^2 + \omega_1^2 - \omega_{12}^2\right) \pm$$

$$\pm \sqrt{\frac{1}{4}\left(\omega_1^2 + \omega_{12}^2 - 2\omega_2^2\right)^2 + \omega_1^2\,\omega_{12}^2}$$

frequency response amplitudes $a_1(\Omega)$ as well as $a_2(\Omega)$ have fixed points independently on d_2 :

$$a_2\left(\Omega_{F_i}\right) = y_{eo} \quad , \quad i = 1,2,3 \quad ,$$

$$a_1\left(\Omega_{F_i}\right) = y_{eo}\,\frac{\omega_1^2}{\left|\omega_1^2 - \Omega_{F_i}^2\right|} \quad , \quad i = 1,2,3 \, .$$

Example 6.2: Centrifuge

At the centrifuge example 2.2, small unsymmetries of the rotor lead to dynamical mass unbalances. The mathematical model of the centrifuge was given by (2.53)

$$\ddot{\phi} + \delta\dot{\phi} + g\Omega\dot{\theta} + k\phi = e\Omega^2\cos\Omega t\,,$$
$$\ddot{\theta} - g\Omega\dot{\phi} + \delta\dot{\theta} + k\theta = e\Omega^2\sin\Omega t$$

with the abbreviations

$$\delta = \frac{d}{I_x}\,,\quad k = \frac{c}{I_x}\,,\quad g = \frac{I_z}{I_x}\,,\quad e = \frac{I_{yz}}{I_x}\,.$$

The frequency response (6.9) reads as

$$q^{(1)} - iq^{(2)} = \frac{e\Omega^2}{\left[k-\Omega^2(1+g)+i\delta\Omega\right]\left[k-\Omega^2(1-g)+i\delta\Omega\right]}\begin{bmatrix} k-\Omega^2(1+g)+i\delta\Omega \\ -i\left[k-\Omega^2(1+g)+i\delta\Omega\right]\end{bmatrix} =$$

$$= \frac{e\Omega^2}{k-\Omega^2(1-g)+i\delta\Omega}\begin{bmatrix}1\\-i\end{bmatrix} \quad \text{if} \quad k-\Omega^2(1+g)+i\delta\Omega \neq 0\,.$$

However, if $k-\Omega^2(1+g)+i\delta\Omega = 0$ the characteristic effect of pseudo-resonance occurs for $\delta = 0$, $\Omega_{R_1}^2 = \dfrac{k}{1+g}$. This is the well-known effect that the counter-rotating eigenmode of centrifuges excited by unbalances does not lead to resonance. On the contrary, the undamped parallel rotating eigenmode of

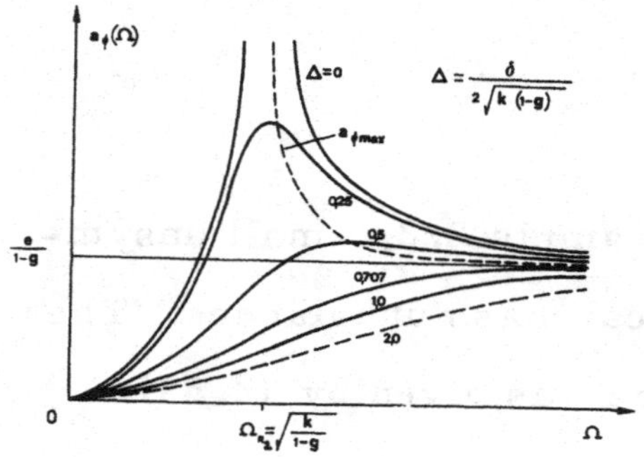

Fig.6.14. Amplitude plot of centrifuge frequency response

the centrifuge yields strict resonance for $\delta=0$, $\Omega_{R_2}^2=\frac{k}{1-g}$. Finally, the amplitude functions are given by

$$a_{\Phi}^2(\Omega)=a_{\theta}^2(\Omega)=$$

$$=\frac{e^2\Omega^4}{\left[k-\Omega^2(1-g)\right]^2+\delta^2\Omega^2}.$$

Typical amplitude plots for various values of the damping are shown in Fig. 6.14.

6.4 Optimization and Absorber Tuning

In the design of vibration systems the parameters have to be found optimally with respect to some criterion. For example, in Fig. 6.13 the amplitude function $a_2(\Omega)$ of the automobile body frequency response is quite different for the various damping coefficients. Also in Fig. 6.14 the amplitude function $a_{\Phi}(\Omega)$ of the centrifuge frequency response depends on the damping ratio Δ. Therefore, the frequency response represents a good tool for an optimal parameter tuning of the vibration system. Therefore, in this section some aspects of parameter optimization with respect to a good frequency response are discussed.

6.4.1 Optimization

Although the meaning of a "good" frequency response is intuitively clear a precise performance criterion is necessary. Usually the goal of optimization is the reduction of the amplitude of the frequency response. ~~While~~ in Fig. 6.14 an increasing damping ratio leads to a decreasing of the amplitudes in Fig. 6.13 the situation is more complicated: in certain frequency domains an increasing damping effects decreasing amplitudes but in other domains an opposite behavior is observed. In applications, the behavior of the amplitudes in Fig. 6.13 is more common than that in Fig. 6.14 which is due to the balance theorem stating that for a certain class of systems the improvement by parameter variations is zero in the mean over all frequencies, see Krebs (1973). Consequently, optimizing the frequency response amplitudes the frequency domain of interest has to be restricted. Then the amplitude functions have to be minimal in the domain $\bar{\Omega} = [\Omega_1, \Omega_2]$ with respect to a performance criterion. Such criteria are discussed in the following.

a) Minimal peak of an amplitude function $a(\Omega)$ on $\bar{\Omega} = [\Omega_1, \Omega_2]$:

$$\max_{\Omega \in [\Omega_1, \Omega_2]} \{a(\Omega)\} \Longrightarrow \text{minimum} . \qquad (6.46)$$

If there are some fixed points in the amplitude plot minimal peaks are often obtained by two optimization steps: (i) put the fixed points as low as possible by variation of those parameters which influence the fixed points, (ii) choose the parameters without influence on the fixed points such that the peaks of the amplitude function coincide with the fixed points (see Klotter (1960)). It has to be noted that this procedure is not always successful. For example, see the amplitude plot of Fig. 6.13. The fixed points are determined as $a_2(\Omega_{F_i}) = y_{eo}$ and $a_2(\omega_1) = 0$ independent on spring or damping parameters Also the peaks can not coincide with the fixed points because $a_2(\Omega_{R_1}) > y_{eo}$ and $a_2(\Omega_{R_2}) > y_{eo}$.

In multi-degree-of-freedom mechanical vibration systems usually more than one amplitude function of $a_1(\Omega),\dots,a_k(\Omega)$ are of interest. Then the performance criterion (6.46) can be replaced by

$$(6.47) \qquad \max_{\Omega \in [\Omega_1, \Omega_2]} \left\{ s_1 a_1(\Omega), \dots, s_k a_k(\Omega) \right\} \Longrightarrow \text{minimum}$$

where s_i, $i = 1(1)k$, are certain weighting factors characterizing the importance of $a_i(\Omega)$.

b) Minimal amplitudes in the mean on $\bar{\Omega} = [\Omega_1, \Omega_2]$:

$$(6.48) \qquad \int_{\Omega_1}^{\Omega_2} a(\Omega)\, d\Omega \Longrightarrow \text{minimum} .$$

In the multidimensional case various generalizations of (6.48) are possible, e.g.

$$\max_{i=1,\dots,k} \left\{ s_i \int_{\Omega_1}^{\Omega_2} a_i(\Omega)\, d\Omega \right\} \Longrightarrow \text{minimum}, \tag{6.49a}$$

$$\sum_{i=1}^{k} s_i \int_{\Omega_1}^{\Omega_2} a_i(\Omega)\, d\Omega \Longrightarrow \text{minimum}. \tag{6.49b}$$

This optimization problem of minimizing the linear area of the amplitude of the frequency response on a certain frequency interval is always well defined but cumbersome in computation.

c) Minimal squared amplitudes in the mean on $\bar{\Omega} = [\Omega_1, \Omega_2]$:

$$\int_{\Omega_1}^{\Omega_2} a^2(\Omega)\, d\Omega \Longrightarrow \text{minimum} \tag{6.50}$$

or

$$\int_{\Omega_1}^{\Omega_2} \sum_{i=1}^{k} s_i\, a_i^2(\Omega)\, d\Omega \Longrightarrow \text{minimum}. \tag{6.51}$$

This performance criterion is equivalent to the requirement

$$\int_{\Omega_1}^{\Omega_2} \mathbf{g}^*(\Omega)\, \mathbf{S}\, \mathbf{g}(\Omega)\, d\Omega \Longrightarrow \text{minimum} \tag{6.52}$$

where $\mathbf{g}(\Omega)$ is the complex vector (5.17) characterizing completely the steady-state frequency response and

$$\mathbf{S} = \mathbf{diag}\,[s_1, \dots, s_n] \tag{6.53}$$

is a symmetric (diagonal) matrix of nonnegative weighting factors s_i.

Using complex curvature integral technique or Fourier transformation technique, respectively, this value of

(6.52) can be calculated in the special case of an infinite frequency domain $\bar{\Omega} = [-\infty, \infty]$. Then

$$\int_{-\infty}^{+\infty} \mathbf{g}^*(\Omega)\,\mathbf{S}\,\mathbf{g}(\Omega)\,d\Omega = \mathbf{f}^* \int_{-\infty}^{+\infty} (-i\Omega\mathbf{E} - \mathbf{A}^T)^{-1}\mathbf{S}(i\Omega\mathbf{E} - \mathbf{A})^{-1} d\Omega\,\mathbf{f} =$$

(6.54)

$$= 2\pi\,\mathbf{f}^* \int_0^{\infty} e^{\mathbf{A}^T t}\,\mathbf{S}\,e^{\mathbf{A}t}\,dt\,\mathbf{f}$$

where $\mathbf{A}$ is the asymptotically stable system matrix and $\mathbf{f}$ is the complex constant input vector of the harmonic excitation vector function. The integral (6.54) can be calculated by the Lyapunov matrix equation (see (B.6) of Appendix B) :

(6.55)
$$\int_{-\infty}^{+\infty} \mathbf{g}^*(\Omega)\,\mathbf{S}\,\mathbf{g}(\Omega)\,d\Omega = 2\pi\,\mathbf{f}^*\mathbf{R}\,\mathbf{f}$$

where $\mathbf{R}$ is the solution of

(6.56)
$$\mathbf{A}^T\mathbf{R} + \mathbf{R}\mathbf{A} = -\mathbf{S}\,.$$

After establishing a performance criterion that will weight the technical requirements effectively, two additional problems exist in optimizing the dynamical behaviour of a vibration system: (i) calculate the criterion, (ii) determine the free parameters such that the criterion takes its optimal value. Both problems can be solved by pencil and paper only for low order systems; for high order systems (usually $n \geq 3$ or 4) calculation and parameter search can only be performed on a digital computer. But these problems are behind the scope of this book and the reader is referred to the proper literature, e.g. Drenick (1967).

6.4.2 Absorber Tuning

A special aspect of optimization with respect to the frequency response is the parameter tuning corresponding to absorption phenomena. As shown in Fig. 6.13 the frequency response of the automobile body is quite satisfactory in the neighborhood of the engine frequency $\omega_1^2 = c_1 / m_1$. The mass m_1 of the engine acts as a vibration absorber of the body motion, i.e. although the excitation force affects the body mass directly, this mass does not vibrate if the excitation frequency is exactly $\Omega = \omega_1$. Then the mass of the engine vibrates in counter-phase to the excitation with an amplitude such that the forces acting on the body mass m_1 are vanishing, i.e. the force due to the first spring (c_1) cancels the force of excitation. This phenomenon is often used to neutralize vibrations of machine foundations by adding a tuned vibration absorber.

The absorption technique is illustrated by an additional example. Consider the automobile wheel suspension of example 6.1. There, the lined amplitude function of the vertical acceleration is reasonably good, see Fig. 6.11. For further improvement of the amplitude func-

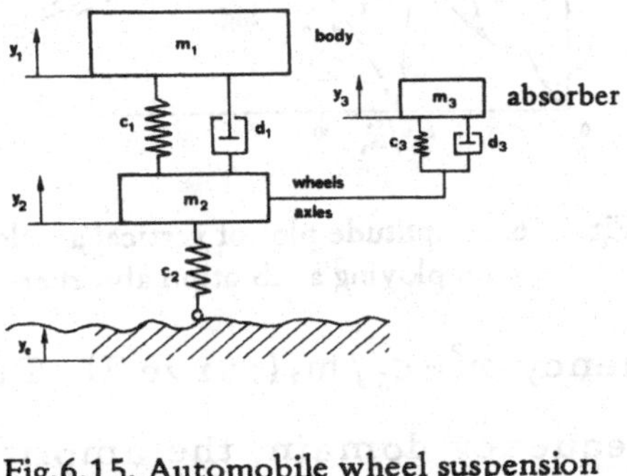

Fig.6.15. Automobile wheel suspension with vibration absorber

tion an absorber has to be used. The mechanical model of the automobile wheel suspension with an absorber mass is shown in Fig. 6.15.

The corresponding equations of motion read as

$$\begin{bmatrix} m_1 & 0 & 0 \\ 0 & m_2 & 0 \\ 0 & 0 & m_3 \end{bmatrix} \begin{bmatrix} \ddot{y}_1 \\ \ddot{y}_2 \\ \ddot{y}_3 \end{bmatrix} + \begin{bmatrix} d_1 & -d_1 & 0 \\ -d_1 & d_1+d_3 & -d_3 \\ 0 & -d_3 & d_3 \end{bmatrix} \begin{bmatrix} \dot{y}_1 \\ \dot{y}_2 \\ \dot{y}_3 \end{bmatrix} +$$

$$+ \begin{bmatrix} c_1 & -c_1 & 0 \\ -c_1 & c_1+c_2+c_3 & -c_3 \\ 0 & -c_3 & c_3 \end{bmatrix} \begin{bmatrix} y_1 \\ y_2 \\ y_3 \end{bmatrix} = \begin{bmatrix} 0 \\ c_2 y_e \\ 0 \end{bmatrix} .$$

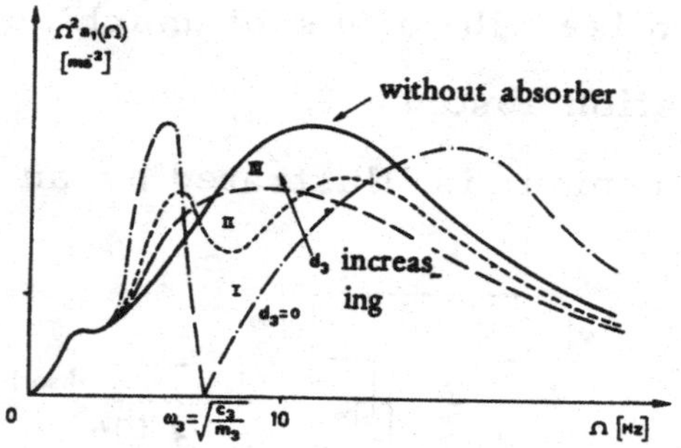

Fig.6.16. Amplitude plot of vertical acceleration employing a vibration absorber

A parameter analysis for the absorber coefficient d_3 results qualitatively in the following amplitude plot of the vertical automobile acceleration, Fig. 6.16. For vanishing absorber damping a complete absorption occurs for the absorber frequency $\omega_3^2 = c_3 / m_3$ (curve I). But in other regions of the interesting frequency domain, the amplitude function is increasing. Therefore, damped absorbers characterized by curves II and III in Fig.

6.16 are preferable featuring a compromise for the whole frequency domain.

CHAPTER 7

Random Vibrations

Mechanical vibration systems are sometimes excited by stochastic forces. Such excitations may appear in all kinds of vibration systems, e.g. an automobile driving on a standard road, a centrifuge with erratic charge or a tall building waving in all weathers. Stochastic forces can not be represented by a single time function, a description by a stochastic process is necessary. Further, the response of the vibration system will be also a stochastic process. Therefore, the stochastic processes will be discussed in short at the beginning.

7.1. Vector Stochastic Processes

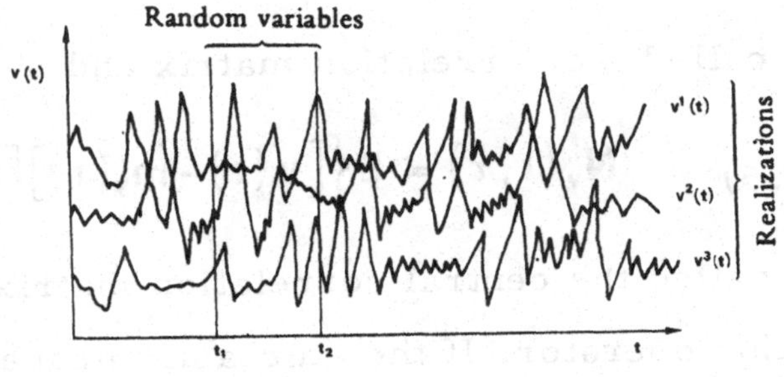

Fig.7.1. Scalar stochastic process

A scalar stochastic process $v(t)$ can be thought of as a family of time function $\{v^{(i)}(t)\}$. Each time is called a realization of the process, Fig.

7.1. For a fixed time the stochastic process is characterized by a random variable. Suppose that $v_i(t), i=1(1)n$, are n scalar stochastic processes which are possibly mutually dependent. Then, it is

$$v(t) = \left[v_1(t)\ v_2(t) \ldots\ldots v_n(t)\right]^T \tag{7.1}$$

a vector stochastic process which can be characterized by the probability distribution

$$\Pr\left\{v(t) : v_i(t_j) \leq v_{ij}\right\}, \quad i = 1(1)n, \quad j = 1(1)m, \tag{7.2}$$

for all v_j, for all $t_j \geq t_0$ and for every number m. In many cases only the first and second-order properties of a stochastic process are important. For the vector stochastic process

$$m_v(t) = E\left\{v(t)\right\} \tag{7.3}$$

is called the mean vector,

$$C_v(t,\tau) = E\left\{v(t)\, v^T(\tau)\right\} \tag{7.4}$$

is called the correlation matrix and

$$N_v(t,\tau) = E\left\{\left[v(t) - m_v(t)\right]\left[v(\tau) - m_v(\tau)\right]^T\right\} \tag{7.5}$$

is called the central correlation matrix where E is the expectation operator. If the stochastic process under consideration has zero mean, $m(t) = 0$, then the correlation matrix and the central correlation matrix coincide. The correlation matrices characterize the coupling of the process at various instants. They have the following properties:

$$\mathbf{C}_v(t,\tau) = \mathbf{C}_v^T(\tau,t),$$

$$\mathbf{N}_v(t,\tau) = \mathbf{N}_v^T(\tau,t) \text{ and} \tag{7.6}$$

$$\mathbf{C}_v(t,\tau) = \mathbf{N}_v(t,\tau) + \mathbf{m}_v(t)\,\mathbf{m}_v^T(\tau)$$

for all t,τ . Since the second-order properties of a stochastic process are equally well characterized by the correlation matrix $\mathbf{C}_v(t,\tau)$ and the central correlation matrix $\mathbf{N}_v(t,\tau)$ usually only $\mathbf{N}_v(t,\tau)$ will be considered.

The covariance matrix $\mathbf{P}_v(t)$ of the stochastic process $\mathbf{v}(t)$ is obtained for $t=\tau$ from the correlation matrix

$$\mathbf{P}_v(t) = \mathbf{N}_v(t,t) \geq 0 \tag{7.7}$$

for all t where $\mathbf{P}_v(t)$ is nonnegative definite. The covariance matrix $\mathbf{P}_v(t)$ summarizes the most essential statistical properties of a stochastic process. In particular, $P_{vii}(t)$ is the variance of the i-th scalar stochastic process $v_i(t)$, $i = 1(1)n$ and $\sigma_{vi}(t) = \sqrt{P_{vii}(t)}$ is the corresponding standard deviation.

A stochastic process $\mathbf{v}(t)$ is called <u>stationary</u> if

$$\Pr\{\mathbf{v}(t) : v_i(t_j) \leq v_{ij}\} = \\ = \Pr\{\mathbf{v}(t+T) : v_i(t_j+T) \leq v_{ij}\} \tag{7.8}$$

for all T . This means that the statistical properties of a stationary process are time-invariant. In particular, it holds

$$(7.9)\qquad \begin{cases} \mathbf{m}_v(t) = \text{const}\,, \\ \mathbf{N}_v(t,\tau) = \mathbf{N}_v(t-\tau)\,, \\ \mathbf{P}_v(t) = \mathbf{P}_v(0) = \text{const}\,, \end{cases}$$

i.e. the correlation matrix depends on $(t-\tau)$ only. For a stationary stochastic process $\mathbf{v}(t)$ the spectral density matrix $\mathbf{S}_v(\Omega)$ can be introduced

$$(7.10)\qquad \mathbf{S}_v(\Omega) = \int_{-\infty}^{+\infty} e^{-i\Omega s}\mathbf{N}_v(s)\,ds$$

where $\mathbf{S}_v(\Omega)$ is defined as the Fourier transform, if it exists, of the correlation matrix $\mathbf{N}_v(t-\tau)$ and $s = t-\tau$. The complex spectral density matrix has the properties

$$(7.11)\qquad \begin{cases} \mathbf{S}_v(-\Omega) = \mathbf{S}_v^T(\Omega) \\ \mathbf{S}_v^*(\Omega) = \mathbf{S}_v(\Omega) \\ \mathbf{S}_v(\Omega) \geq 0 \end{cases}$$

for all Ω . Further, it yields

$$\mathbf{N}_v(s) = \frac{1}{2\pi} \int_{-\infty}^{\infty} \mathbf{S}_v(\Omega) e^{i\Omega s} d\Omega . \tag{7.12}$$

The spectral density matrix is a successful mean to obtain experimentally data of a stationary stochastic process.
A stochastic process $\mathbf{v}(t)$ is called Gaussian if for each set of instants of time t_j, $j = 1(1)m$, the $n \times 1$ -random vector $\mathbf{v}(t_j)$ has a Gaussian probability distribution. Since the Gaussian probability distribution is completely characterized by the first and second-order properties, a Gaussian stochastic process $\mathbf{v}(t)$ is completely characterized by the mean vector $\mathbf{m}_v(t)$ and the correlation matrix $\mathbf{N}_v(t, \tau)$.
A stochastic process $\mathbf{v}(t)$ is called a Markov process if

$$\begin{aligned} &Pr\left\{v_i(t_\ell) \leq v_{i\ell} \,\middle|\, v_i(t_1), v_i(t_2), \ldots v_i(t_{\ell-1})\right\} = \\ &\qquad Pr\left\{v_i(t_\ell) \leq v_{i\ell} \,\middle|\, v_i(t_{\ell-1})\right\}, \; i = 1(1)n \end{aligned} \tag{7.13}$$

for all l, for all $t_1, t_2, \ldots, t_l$ with $t_l \geq t_{l-1} \geq \ldots \geq t_1$ and for all $\mathbf{v}_{il}$. The symbol $Pr\{A\,|\,B\}$ means conditional probability, i.e. the probability of A if it is known that B already occurred. The Markov process does not depend on the events in the past.

A stochastic process $\mathbf{v}(t)$, $t \geq t_0$ with independent increments is given if

$$(7.14)\quad \begin{cases} v(t_0) = 0 \,, \\ E\{v(t_2)-v(t_1)\} = E\{v(t_4)-v(t_3)\} = 0 \,, \\ E\{[v(t_2)-v(t_1)][v(t_4)-v(t_3)]^T\} = 0 \end{cases}$$

for any sequence of instants t_1, t_2, t_3, t_4 with $t_0 \leq t_1 \leq \leq t_2 \leq t_3 \leq t_4$. A stochastic process $\mathbf{v}(t)$ is called a Wiener process if it is a process with independent increments where each of the increments $[\mathbf{v}(t_2) - \mathbf{v}(t_1)]$ is a Gaussian random vector with zero mean and covariance matrix $\mathbf{P} = \mathbf{Q}(t_2 - t_1)$ where $\mathbf{Q}$ is a constant, nonnegative intensity matrix.

A stochastic process $\mathbf{v}(t)$ is called a stationary white noise process if it is an idealized Gauss-Markov process with independent increments even for $(t_2 - t_1) \longrightarrow 0$. The properties of such a process are given by

$$(7.15)\quad \begin{cases} \mathbf{m}_v(t) = 0 \\ \mathbf{N}_v(t,\tau) = \mathbf{Q}_v \delta(t-\tau) \,, \; \mathbf{Q}_v \geq 0 \end{cases}$$

where $\mathbf{Q}_v$ is the constant, nonnegative definite intensity matrix of the white noise process and $\delta(t-\tau)$ is the Dirac function (5.2). The corresponding spectral density matrix of this process follows from (7.10) and (7.15) as

$$\mathbf{S}_v(\Omega) = \mathbf{Q}_v \,. \tag{7.16}$$

Thus, the stationary white noise process is completely given by the intensity matrix $\mathbf{Q}_v$. Due to the small correlation even between two near values $\mathbf{v}(t_1)$ and $\mathbf{v}(t_2)$ the white noise process is very irregular and contains signals at quite high frequencies. Therefore, the covariance matrix of the white noise process has an infinite value

$$\mathbf{P}_v(t) = \mathbf{Q}_v\,\delta(0) \rightarrow \infty \tag{7.17}$$

which immediately points out that this process does not exist in the physical world. However, if the white noise process has passed an integration, one obtains the Wiener process and is again on a firm physical background. In the following sections the response of vibration systems to stochastic excitations is considered. But this means at least one integration and, therefore, the white noise process can be applied successfully for the excitation. For a more extensive discussion of stochastic processes in theory and application consult, e.g., Papoulis (1965), Parkus (1969) Jazwinski (1970) Kwakernaak and Sivan (1972).

7.2 Response to Stochastic Forces

The stochastic forces acting on a vibration system may be modeled by a stationary white noise process characterized by

$$\mathbf{m}_f(t) = 0\,, \qquad \mathbf{N}_f(t,\tau) = \mathbf{Q}\,\delta(t-\tau) \tag{7.18}$$

where $\mathbf{Q}$ is the $n \times n$ -intensity matrix. Due to the stochastic excitation, the vibration system is now governed by a stochastic differential equation. But for the white noise excitation formally the deterministic differential equation

$$\dot{\mathbf{x}}(t) = \mathbf{A}\mathbf{x}(t) + \mathbf{f}(t)\,, \quad \mathbf{x}(0) = \mathbf{x}_0 \tag{7.19}$$

can be used as well-known from control theory, e.g. Jazwinski (1970). However, the initial state $\mathbf{x}_0$ has to be a Gaussian random vector

$$\mathbf{x}_0 \sim (\mathbf{m}_0\,, \mathbf{P}_0) \tag{7.20}$$

independent of the white noise excitation (7.18), where $\mathbf{m}_0$ is the $n \times 1$ -mean vector and $\mathbf{P}_0$ is the $n \times n$ -covariance matrix. Then, the general solution is given by

$$\mathbf{x}(t) = \boldsymbol{\Phi}(t)\mathbf{x}_0 + \int_0^t \boldsymbol{\Phi}(t-\tau)\,\mathbf{f}(\tau)\,d\tau \tag{7.21}$$

where $\Phi(t)$ is the fundamental matrix (3.7) and the integral is a stochastic integral. It can be shown that the stochastic response $x(t)$ is a nonstationary Gauss-Markov process. Such a process is completely specified by the knowledge of the mean vector $m_x(t)$ and the correlation matrix $N_x(t_1, t_2)$. For the further evaluation of (7.21), two properties of the stochastic integral introduced by Ito (1944) are presented:

$$E \int_{t_0}^{t} \Phi(t-\tau) f(\tau) d\tau = 0 , \tag{7.22}$$

$$E\left\{\left[\int_{t_0}^{t_1} \Phi(t_1-\tau_1) f(\tau_1) d\tau_1\right]\left[\int_{t_0}^{t_2} \Phi(t_2-\tau_2) f(\tau_2) d\tau_2\right]^T\right\} =$$

$$= \int_{t_0}^{t_1}\int_{t_0}^{t_2} \Phi(t_1-\tau_1) Q \delta(\tau_2-\tau_1) \Phi^T(t_2-\tau_2) d\tau_1 d\tau_2 =$$

$$= \int_{t_0}^{\min(t_1,t_2)} \Phi(t_1-\tau) Q \Phi^T(t_2-\tau) d\tau . \tag{7.23}$$

Then, it holds

$$m_x(t) = \Phi(t) m_0 \tag{7.24}$$

for the mean vector and

$$N_x(t_1, t_2) = \Phi(t_1) P_0 \Phi^T(t_2) + \\ + \int_{t_0}^{\min(t_1,t_2)} \Phi(t_1-\tau) Q \Phi^T(t_2-\tau) d\tau \tag{7.25}$$

for the correlation matrix. Further, the covariance matrix $\mathbf{P}_x(t) = \mathbf{N}_x(t, t)$ is easily obtained from (7.25) by $t_1 = t_2 = t$ and $t_0 = 0$ as

$$(7.26) \qquad \mathbf{P}_x(t) = \Phi(t)\mathbf{P}_0\,\Phi^T(t) + \int_0^t \Phi(t-\tau)\,\mathbf{Q}\,\Phi^T(t-\tau)\,d\tau .$$

Thus, the covariance matrix $\mathbf{P}_x(t)$ of the response $\mathbf{x}(t)$ is available if the integral in (7.26) is solved. Introducing (3.7) the integral reads as

$$(7.27) \qquad \int_0^t e^{\mathbf{A}(t-\tau)}\mathbf{Q}\,e^{\mathbf{A}^T(t-\tau)}\,d\tau = e^{\mathbf{A}t}\int_0^t e^{-\mathbf{A}\tau}\mathbf{Q}\,e^{-\mathbf{A}^T\tau}\,d\tau\; e^{\mathbf{A}^Tt}$$

and can be finally solved by

$$(7.28) \qquad \int_0^t e^{-\mathbf{A}\tau}\mathbf{Q}\,e^{-\mathbf{A}^T\tau}\,d\tau = e^{-\mathbf{A}t}\mathbf{P}e^{-\mathbf{A}^Tt} - \mathbf{P}$$

where $\mathbf{P}$ is a constant $n \times n$ -matrix following from the Lyapunov matrix equation

$$(7.29) \qquad \mathbf{A}\mathbf{P} + \mathbf{P}\mathbf{A}^T + \mathbf{Q} = \mathbf{0} .$$

The integral (7.28) is plausible by differentiation and it can be proved by the series expansion (3.7) of the fundamental matrix $e^{\mathbf{A}t}$. The conditions for existence and uniqueness of the matrix $\mathbf{P}$, and therefore for the integral (7.28) are given in Appendix B.

The first and second-order statistical properties of the stochastic response $\mathbf{x}(t)$ can now be summarized as

$$\begin{aligned} \mathbf{m}_x(t) &= \boldsymbol{\Phi}(t)\,\mathbf{m}_0 \\ \mathbf{P}_x(t) &= \boldsymbol{\Phi}(t)\,(\mathbf{P}_0 - \mathbf{P})\,\boldsymbol{\Phi}^T(t) + \mathbf{P}\,. \end{aligned} \tag{7.30}$$

In particular, $P_{ii}(t)$ offers the variance and $\sigma_i(t) = \sqrt{P_{ii}(t)}$ the standard deviation of the i-th state variable $x_i(t)$. The steady-state stochastic response is obtained uniquely for asymptotically stable systems as a consequence of Theorem B.2. In particular, it holds

$$\mathbf{m}_x(t \to \infty) = \mathbf{0}\,, \tag{7.31}$$

$$\mathbf{P}_x(t \to \infty) = \mathbf{P} = \int_0^\infty e^{\mathbf{A}\tau}\,\mathbf{Q}\,e^{\mathbf{A}^T\tau}\,d\tau = \text{const} \tag{7.32}$$

if (7.19) is asymptotically stable. Thus, the covariance matrix $\mathbf{P}$ is the most essential part of the steady-state response. It can be computed via the covariance analysis or via the spectral density analysis as shown in the next sections.

7.3 Covariance Analysis

The covariance analysis reduces the computation of the stochastic response to the solution of the algebraic Lyapunov equation (7.29). In Appendix B four different solution methods for the Lyapunov equation are outlined. Here, the explicit solution, Theorem B.4, and the implicit solution, Theorem B.5, will be presented with some modifications.

For the explicit solution, formula (B.9) can be rewritten as

$$P = \frac{1}{2a_0 \det \mathbf{H}} \sum_{k=0}^{n-1} H_{k+1,1} \sum_{m=0}^{2k} (-1)^m \mathbf{A}_m \mathbf{Q} \mathbf{A}_{2k-m}^T \tag{7.33}$$

where

$$\mathbf{A}_m = \mathbf{A}\mathbf{A}_{m-1} + a_m \mathbf{E} \tag{7.34}$$

is a $n \times n$ -matrix, a_m the m-th characteristic coefficient, $\mathbf{H}$ the $n \times n$ -Hurwitz matrix (4.14) and $H_{K+1,1}$ the cofactor of the $k+1,1$-th element of $\mathbf{H}$. This formula is extremely useful for the analytical investigations of low order systems as shown by Schiehlen (1973 a).

The implicit solution, Theorem B.5, can be simplified for the following linear mechanical system

$$\mathbf{M}\ddot{\mathbf{y}}(t) + (\mathbf{D}+\mathbf{G})\dot{\mathbf{y}}(t) + (\mathbf{K}+\mathbf{N})\mathbf{y}(t) = \mathbf{h}(t) \tag{7.35}$$

where $\mathbf{h}(t)$ is the $f \times 1$ -stationary white noise vector process with the $f \times f$ -intensity matrix $\mathbf{V}$. Then, the first order system (7.19) is characterized by

$$x = \begin{bmatrix} y \\ \dot{y} \end{bmatrix}, \quad A = \begin{bmatrix} O & E \\ -M^{-1}(K+N) & -M^{-1}(D+G) \end{bmatrix},$$

$$f = \begin{bmatrix} O \\ M^{-1}h \end{bmatrix}, \quad Q = \begin{bmatrix} O & O \\ O & M^{-1}VM^{-1} \end{bmatrix} \qquad (7.36)$$

where f is the $n \times 1$ -stationary white noise vector process with the $n \times n$ -intensity matrix Q, $n = 2f$. Further, the $n \times n$-covairance matrix P is divided in $f \times f$ -submatrices P_I P_{II}, P_{III} :

$$P = \begin{bmatrix} P_I & P_{II} \\ P_{II}^T & P_{III} \end{bmatrix} \qquad (7.37)$$

Then, using A, Q and P from (7.36) and (7.37) the Lyapunov equation (7.29) can be extended as

$$\left.\begin{aligned} & P_{II} + P_{II}^T = O, \\ & (K+N)P_I + (D+G)P_{II}^T - MP_{III} = O, \\ & (K+N)P_{II}M + (D+G)P_{III}M + \\ & + MP_{II}^T(K-N) + MP_{III}(D-G) - V = O. \end{aligned}\right\} \qquad (7.38)$$

This means that the submatrix P_{II} is a skew-symmetric matrix and $n/4\,(n/2+1)$ essential elements are cancelled within the covariance matrix P. Thus, in the Lyapunov equation (7.29) or (7.38), respectively, remain only $n/4\,(3n/2+1)$ unknown variables and it can be rewritten as a common linear equation

(7.39) $$\mathcal{A}\,\mathbf{r} = \mathbf{s}$$

with the reduced order of $n/4\,(3n/2+1)$. The implicit solution presented in this section is often restricted to low order systems for numerical reasons. But even if other methods for the computation of the Lyapunov equation of mechanical vibration systems is used, the skew-symmetry of the submatrix $\mathbf{P}_{II}$ is maintained and can be used to check the numerical results, see also Schiehlen (1974).

7.4 Spectral Density Analysis

The spectral density analysis, widely applied in random vibrations, uses the simple algebraic formula for the spectral density of the solution process and the inverse Fourier transform to obtain the covariance matrix. From (7.23) and (7.25), it follows for the steady-state correlation matrix of the solution process $\mathbf{x}(t\to\infty)$ of an asymptotically stable system

(7.40) $$\mathbf{N}_x(s) = \int_{-\infty}^{\infty}\int_{-\infty}^{\infty} \boldsymbol{\Phi}(z_1)\,\mathbf{Q}\,\delta(s+z_2-z_1)\,\boldsymbol{\Phi}^T(z_2)\,dz_1\,dz_2$$

where $s = t_1 - t_2$, $z_1 = t_1 - \tau_1$, $z_2 = t_2 - \tau_2$. The Fourier transform $S_x(\Omega)$ of (7.40) can now be calculated by (7.10) resulting in

$$\mathbf{S}_x(\Omega) = \mathbf{F}(\Omega)\,\mathbf{Q}\,\mathbf{F}^T(-\Omega) \tag{7.41}$$

where $\mathbf{F}(\Omega)$ is the frequency response matrix (5.18) representing also the Fourier transform of the transition matrix $\boldsymbol{\Phi}(t)$. Therefore, the spectral density matrix $\mathbf{S}_x(\Omega)$ of the solution process $\mathbf{x}(t)$ can be easily obtained. But the spectral density is of slight interest only. In technical applications the covariance matrix $\mathbf{P}$ is essential. The matrix $\mathbf{P}$ follows from (7.12) for $s = 0$:

$$\mathbf{P} = \frac{1}{2\pi}\int_{-\infty}^{\infty} \mathbf{S}_x(\Omega)\, d\Omega . \tag{7.42}$$

Thus, an infinite integral has to be solved for each element of the spectral density matrix $\mathbf{S}_x(\Omega)$. This is a hard job. For the standard deviations σ_i of the variances P_{ii}, the infinite integrals have been solved by James, Nichols and Phillips (1947) in the early days of stochastic control theory :

$$\sigma_i^2 = \frac{1}{2\pi}\int_{-\infty}^{\infty} S_x(\Omega)_{ii}\, d\Omega = \frac{1}{2\pi i}\int_{-\infty}^{\infty} \frac{g_n(\Omega)}{h_n(\Omega)h_n(-\Omega)}\, d\Omega \tag{7.43}$$

where

$$\left.\begin{aligned} g_n(\Omega) &= b_0\,\Omega^{2n-2} + b_1\,\Omega^{2n-4} + \ldots + b_{n-1}\,, \\ h_n(\Omega) &= a_0(i\Omega)^n + a_1(i\Omega)^{n-1} + \ldots + a_n = \det(i\Omega\mathbf{E} - \mathbf{A}) \end{aligned}\right\} \tag{7.44}$$

are polynomials in Ω and, in particular, $h_n(\Omega)$ represents the characteristic equation of the vibration system. However, the integrals (7.43) have been completely evaluated only for systems with order $n \leq 7$. Further, these integrals have been found by Crandall and Mark (1963), Fabian (1973) and some of the other authors dealing with random vibrations. Thus, the integrals do not have to be repeated here.

Comparing the covariance analysis and the spectral density analysis it turns out that the covariance analysis is more adequate for the solution of the random vibration problems. The spectral density analysis, however, is more popular for historical reasons. The filter theory, strongly related with random vibration theory, has been firstly developed in the frequency domain by Wiener (1949) while the extension to the time domain was given later by Kalman (1960). However, in random vibrations the development from spectral density analysis to the covariance analysis is going on very slowly. For integrity, it has to be mentioned that the spectral density analysis is restricted to the steady-state response of time-invariant systems while the covariance analysis allows the investigation of the transition response as well as the response of time-variant systems.

7.5 Modeling of Stochastic Processes

The white noise is a very successful mean to overcome mathematical problems related to stochastic differential equations. On the other hand, the white noise process does not exist in the physical world and therefore the intensity can never be exactly measured. However, for stationary stochastic processes, the intensity can be approximated very closely.

Assume the spectral density $S_v(\Omega)$ of the real scalar process $v(t)$ as shown in Fig. 7.2.

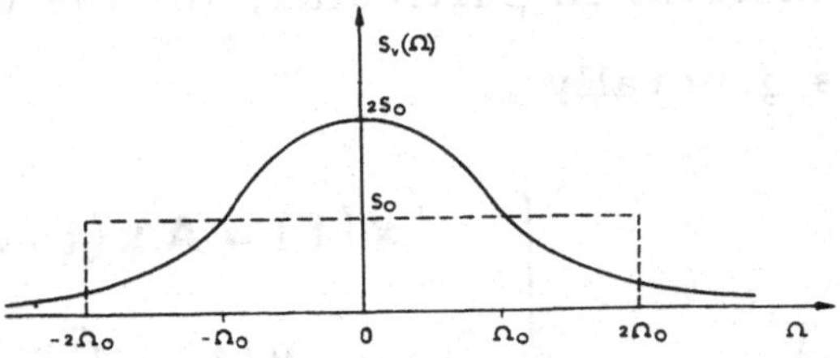

Fig.7.2. Typical spectral density of a real process

Then, three cases may be considered:

1. Slow system. If $\max|\lambda_i| << \Omega_0, i=1(1)n$, then $Q_v = 2\,S_0$ where λ_i is the i-th eigenvalue of the system under consideration and Ω_0, S_0 follow from Fig. 7.2.
2. Medium system. If $\max|\lambda_i| \approx \Omega_0$, $i=1(1)n$, then $Q_v = S_0$ or $Q_v = \sigma_v^2/2\Omega_0$ where λ_i is the i-th eigenvalue of the system, Ω_0, S_0 follow from Fig. 7.2. and σ_v is the standard deviation of the real process $v(t)$.
3. Fast system. If $\max|\lambda_i| >> \Omega_0$, then the dynamic modeling of the colored noise $v(t)$ is necessary. It yields

$$\dot{v}(t) = a v(t) + b w(t) \tag{7.45}$$

where a , b are constants and $w(t)$ is a white noise process with intensity Q_W . Then, by the constants a, b the spectral density $S_V(\Omega)$ can be approximated closely. For analysis, the system (7.19) has to be completed by (7.45), often called colored noise shaping filter.

For a vector process, these considerations have to be applied to each diagonal element of the spectral density matrix. In particular, for the colored noise modeling, it holds generally

$$\left\{\begin{aligned} \dot{\mathbf{x}}(t) &= \mathbf{A}\,\mathbf{x}(t) + \mathbf{v}(t), \\ \mathbf{v}(t) &= \bar{\mathbf{C}}\,\boldsymbol{\xi}(t), \\ \dot{\boldsymbol{\xi}}(t) &= \bar{\mathbf{A}}\,\boldsymbol{\xi}(t) + \bar{\mathbf{B}}\,\mathbf{w}(t) \end{aligned}\right. \tag{7.46}$$

or summarized

$$\begin{bmatrix} \dot{\mathbf{x}} \\ \dot{\boldsymbol{\xi}} \end{bmatrix} = \begin{bmatrix} \mathbf{A} & \bar{\mathbf{C}} \\ 0 & \bar{\mathbf{A}} \end{bmatrix} \begin{bmatrix} \mathbf{x} \\ \boldsymbol{\xi} \end{bmatrix} + \begin{bmatrix} 0 \\ \bar{\mathbf{B}} \end{bmatrix} \mathbf{w}(t) \tag{7.47}$$

where $\mathbf{w}(t)$ is stationary white noise, $\mathbf{v}(t)$ colored white noise and $\bar{\mathbf{A}}$, $\bar{\mathbf{B}}$, $\bar{\mathbf{C}}$ are matrices of corresponding dimensions, available for the modeling of colored noise. Obviously, the order of the system may be increasing strongly. But the solution approach for (7.19) presented in section 7.2 suits for (7.47) just so well.

7.6. Examples

An automobile driving on a standard road and a centrifuge with erratic charge will be used as examples. The automobile problem is treated by covariance and spectral density analysis while the centrifuge is investigated by the explicit and implicit solution available for the covariance analysis.

Example 7.1: Automobile

An automobile will be driving with constant velocity v on a standard road given by $y_e(t)$, Fig.7.3. The automobile is modeled by a body with mass m and a linear suspension with spring coefficient c and damping coefficient d. The masses of wheels and axles are neglected. Then, the equation for the vertical motion $y(t)$ reads as

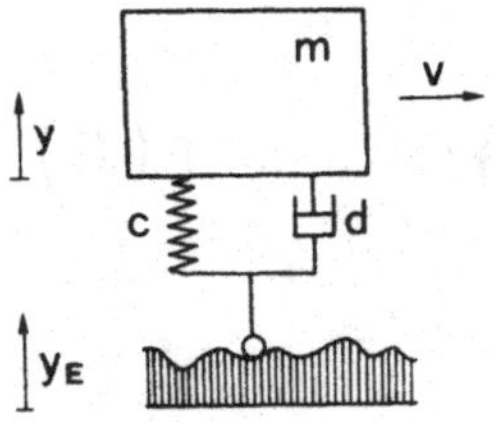

Fig.7.3. Automobile model

$$\ddot{y}(t) + D\dot{y}(t) + Ky(t) = h(t) \tag{7.48}$$

where $D = d/m, K = c/m$ and $h(t) = D\dot{y}_e(t) + Ky(t)$ is a scalar white noise process with intensity V. The corresponding first order system is given by

(7.49) $$\dot{\mathbf{x}}(t) = \mathbf{A}\mathbf{x}(t) + \mathbf{f}(t)$$

where

(7.50) $$\begin{cases} \mathbf{x} = \begin{bmatrix} y \\ \dot{y} \end{bmatrix}, & \mathbf{A} = \begin{bmatrix} 0 & 1 \\ -K & -D \end{bmatrix}, \\ \mathbf{f} = \begin{bmatrix} 0 \\ h \end{bmatrix}, & \mathbf{Q} = \begin{bmatrix} 0 & 0 \\ 0 & V \end{bmatrix}. \end{cases}$$

Due to theorem 4.13, system (7.48) or (7.49), respectively, is asymptotically stable for $c > 0, d > 0$.

Firstly, the spectral density analysis is applied. The frequency response matrix (5.18) follows as

(7.51) $$F(\Omega) = \frac{1}{K - \Omega^2 + i\Omega D} \begin{bmatrix} i\Omega + D & 1 \\ -K & i\Omega \end{bmatrix}.$$

The spectral density matrix of the solution process results from (7.41)

$$S_x(\Omega) = \frac{1}{(K-\Omega^2)^2 + (D\Omega)^2} \begin{bmatrix} i\Omega + D & 1 \\ -K & i\Omega \end{bmatrix} \begin{bmatrix} 0 & 0 \\ 0 & V \end{bmatrix} \begin{bmatrix} -i\Omega + D & -K \\ 1 & -i\Omega \end{bmatrix} =$$

(7.52)

$$= \frac{V}{(K-\Omega^2)^2 + (D\Omega)^2} \begin{bmatrix} 1 & -i\Omega \\ i\Omega & \Omega^2 \end{bmatrix}$$

Then the covariance matrix can be found from (7.42)

$$P = \frac{V}{2\pi} \int_{-\infty}^{\infty} \begin{bmatrix} 1 & -i\Omega \\ i\Omega & \Omega^2 \end{bmatrix} \frac{d\Omega}{(K-\Omega^2)^2 + (D\Omega)^2} . \qquad (7.53)$$

The solution of the integral (7.53) will be presented for the variances. It holds

$$P_{11} = \frac{V}{2\pi i} \int_{-\infty}^{\infty} \frac{i\, d\Omega}{(-\Omega^2 + i\Omega D + K)(-\Omega^2 - i\Omega D + K)} , \qquad (7.54)$$

$$P_{22} = \frac{V}{2\pi i} \int_{-\infty}^{\infty} \frac{i\Omega^2 d\Omega}{(-\Omega^2 + i\Omega D + K)(-\Omega^2 - i\Omega D + K)} . \qquad (7.55)$$

Now, the typical integrals (7.43) are obtained. From James, Nichols and Phillips (1947) it follows the result

$$\frac{1}{2\pi i} \int_{-\infty}^{\infty} \frac{b_0 \Omega^2 + b_1}{(a_0 \Omega^2 + a_1 \Omega + a_2)(a_0 \Omega^2 - a_1 \Omega + a_2)} d\Omega = \frac{a_0 b_1 - a_2 b_0}{2 a_0 a_1 a_2} .$$

(7.56)

By comparison of (7.56) with (7.54) or (7.55), one gets the coefficients $a_0 = -1$, $a_1 = iD$, $a_2 = K$ and $b_0 = 0$, $b_1 = i$ or $b_0 = i$, $b_1 = 0$. Then, the result is finally obtained by (7.56):

$$\sigma_y^2 = P_{11} = V / 2DK, \qquad (7.57)$$

$$\sigma_{\dot{y}}^2 = P_{22} = V / 2D. \qquad (7.58)$$

Secondly, the covariance analysis is used. The Lyapunov equation (7. 29) can be expanded immediately:

$$\begin{bmatrix} 0 & 1 \\ -K & -D \end{bmatrix}\begin{bmatrix} P_{11} & P_{12} \\ P_{12} & P_{22} \end{bmatrix} + \begin{bmatrix} P_{11} & P_{12} \\ P_{12} & P_{22} \end{bmatrix}\begin{bmatrix} 0 & -K \\ 1 & -D \end{bmatrix} = \begin{bmatrix} 0 & 0 \\ 0 & -V \end{bmatrix} \tag{7.59}$$

or

$$2\,P_{12} = 0, \tag{7.60}$$

$$D\,P_{12} + K\,P_{11} - P_{22} = 0, \tag{7.61}$$

$$2\,K P_{12} + 2\,D P_{22} = V. \tag{7.62}$$

Thus, from (7.62) it follows (7.58), from (7.61) and (7.58) one gets (7.57) and (7.60) verifies the skew symmetry of the submatrix P_{12} which means zero for $f = 1$. This simple example shows obviously the advantage of the covariance analysis.

Example 7.2: Centrifuge

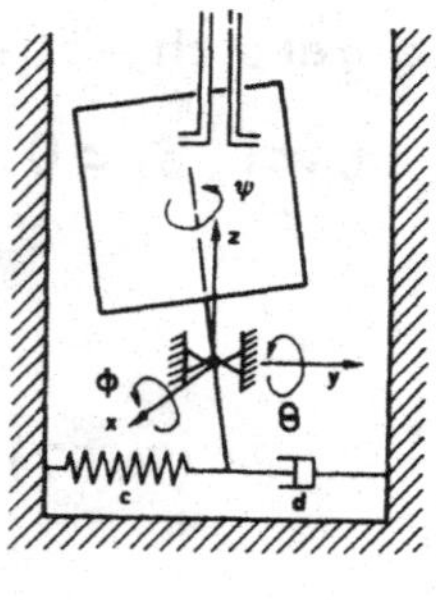

Fig.7.4. Centrifuge model

A centrifuge is charged during the operation time with liquid blowing stochastically out of two pipes, Fig.7.4. The centrifuge is modeled as a symmetric, rigid body with moments of inertia I_x, I_z . The centrifuge will be spinning with constant angular velocity $\dot{\psi}$ and there is an elastic, symmetric suspension characterized by c and d. Then, the equation of motion for the small angles $\phi(t), \theta(t)$ reads as

$$\ddot{\mathbf{y}}(t) + (\mathbf{D}+\mathbf{G})\dot{\mathbf{y}}(t) + \mathbf{K}\mathbf{y}(t) = \mathbf{h}(t) \qquad (7.63)$$

where $\mathbf{y}=[\phi\ \theta]^T$, $\mathbf{D}=\delta\mathbf{E}$, $\mathbf{G}=\omega\mathbf{S}$, $\mathbf{K}=k\mathbf{E}$ and $\delta=d/I_x$, $\omega=I_z\dot{\psi}/I_x$, $k=c/I_x$. Further, $\mathbf{h}(t)$ is a white noise process with intensity matrix

$$\mathbf{V} = \begin{bmatrix} \nu & 0 \\ 0 & 0 \end{bmatrix}, \qquad (7.64)$$

i.e., the stochastic forces are acting only in the ϕ-direction. The corresponding first order system (7.19) is characterized by the matrices

$$\mathbf{A} = \left[\begin{array}{cc|cc} 0 & 0 & 1 & 0 \\ 0 & 0 & 0 & 1 \\ \hline -k & 0 & -\delta & -\omega \\ 0 & -k & \omega & -\delta \end{array}\right], \quad \mathbf{Q} = \left[\begin{array}{cc|cc} 0 & 0 & 0 & 0 \\ 0 & 0 & 0 & 0 \\ \hline 0 & 0 & \nu & 0 \\ 0 & 0 & 0 & 0 \end{array}\right]. \qquad (7.65)$$

Due to Theorem 4.14, the system (7.63) is asymptotically stable for $\delta>0, k>0$.

Firstly the explicit solution is applied. From (7.33), it follows

$$\mathbf{P} = \frac{1}{2a_0 \det \mathbf{H}} \sum_{k=0}^{3} H_{k+1,1} \sum_{m=0}^{6} (-1)^m \mathbf{A}_m \mathbf{Q} \mathbf{A}_{2k-m}^T . \qquad (7.66)$$

The matrices $\mathbf{A}_m$ are according to (7.34) given by

(7.67) $$\mathbf{A}_0 = \mathbf{E},$$

(7.68) $$\mathbf{A}_1 = \begin{bmatrix} 2\delta & 0 & 1 & 0 \\ 0 & 2\delta & 0 & 1 \\ -k & 0 & \delta & -\omega \\ 0 & -k & \omega & \delta \end{bmatrix},$$

(7.69) $$\mathbf{A}_2 = \begin{bmatrix} k+\delta^2+\omega^2 & 0 & \delta & -\omega \\ 0 & k+\delta^2+\omega^2 & \omega & \delta \\ -k\delta & k\omega & k & 0 \\ -k\omega & -k\delta & 0 & k \end{bmatrix},$$

(7.70) $$\mathbf{A}_3 = \begin{bmatrix} k\delta & k\omega & k & 0 \\ -k\omega & k\delta & 0 & k \\ -k^2 & 0 & 0 & 0 \\ 0 & -k^2 & 0 & 0 \end{bmatrix},$$

(7.71) $$\mathbf{A}_4 = \mathbf{A}_5 = \mathbf{A}_6 = 0 .$$

The coefficients a_m of the characteristic equation read as

(7.72) $$\begin{cases} a_0 = 1, \quad a_1 = 2\delta, \quad a_2 = 2k+\delta^2+\omega^2, \\ a_3 = 2\delta k, \quad a_4 = k^2, \quad a_5 = a_6 = 0, \end{cases}$$

and the Hurwitz matrix (4.14) is for

$$\mathbf{H} = \begin{bmatrix} a_1 & 1 & 0 & 0 \\ a_3 & a_2 & a_1 & 1 \\ 0 & a_4 & a_3 & a_2 \\ 0 & 0 & 0 & a_4 \end{bmatrix} \tag{7.73}$$

Then, the cofactors are obtained

$$\left.\begin{aligned} H_{11} &= 2\delta k^3 (k + \delta^2 + \omega^2), \\ H_{21} &= -2\delta k^3, \\ H_{31} &= 2\delta k^2, \\ H_{41} &= -2\delta (k + \delta^2 + \omega^2) \end{aligned}\right\} \tag{7.74}$$

and the Hurwitz determinant follows as

$$\det \mathbf{H} = a_1 H_{11} + a_3 H_{21} = 4\delta^2 k^3 (\delta^2 + \omega^2). \tag{7.75}$$

Introducing (7.67) ÷ (7.75) in (7.66) one gets immediately the covariance matrix

$$\mathbf{P} = \frac{\nu}{4\delta k(\delta^2 + \omega^2)} \begin{bmatrix} 2\delta^2+\omega^2 & \delta\omega & 0 & 0 \\ \delta\omega & \omega^2 & 0 & 0 \\ 0 & 0 & (2\delta^2+\omega^2)k & \delta\omega k \\ 0 & 0 & \delta\omega k & \omega^2 k \end{bmatrix}. \tag{7.76}$$

The submatrices $\mathbf{P}_I$ and $\mathbf{P}_{III}$ are symmetric while the submatrix $\mathbf{P}_{II}$ is zero which is not a contradiction to the skew-symmetry predicted in section 7.3. The spectral densities $\sigma_\Phi, \sigma_{\dot\Phi}, \sigma_\Theta, \sigma_{\dot\Theta}$ are the square roots of the elements P_{11}, P_{22}, P_{33}, P_{44}. They are not listed here.

Secondly, the implicit solution is used. The submatrices

$$\mathbf{P}_I = \begin{bmatrix} P_{11} & P_{12} \\ P_{12} & P_{22} \end{bmatrix}, \quad \mathbf{P}_{II} = \begin{bmatrix} 0 & P_{14} \\ -P_{14} & 0 \end{bmatrix}, \quad \mathbf{P}_{III} = \begin{bmatrix} P_{33} & P_{34} \\ P_{34} & P_{44} \end{bmatrix} \tag{7.77}$$

indicate that the covariance matrix has $4/4\,(3\cdot4/2+1) = 7$ essential elements. Then, the Lyapunov equation (7.21) can be converted in an ordinary equation system of order 7:

$$\left[\begin{array}{cccc|ccc} k & 0 & 0 & \omega & -1 & 0 & 0 \\ 0 & k & 0 & -\delta & 0 & -1 & 0 \\ 0 & k & 0 & \delta & 0 & -1 & 0 \\ 0 & 0 & k & \omega & 0 & 0 & -1 \\ \hline 0 & 0 & 0 & 0 & 2\delta & 2\omega & 0 \\ 0 & 0 & 0 & 0 & -\omega & 2\delta & \omega \\ 0 & 0 & 0 & 0 & 0 & -2\omega & 2\delta \end{array}\right] \left[\begin{array}{c} P_{11} \\ P_{12} \\ P_{22} \\ P_{14} \\ \hline P_{33} \\ P_{34} \\ P_{44} \end{array}\right] = \left[\begin{array}{c} 0 \\ 0 \\ 0 \\ 0 \\ \hline \nu \\ 0 \\ 0 \end{array}\right]. \tag{7.78}$$

The solution of the linear equation system (7.78) can be presented analytically

$$\begin{bmatrix} P_{11} \\ P_{12} \\ P_{22} \\ P_{14} \\ \hdashline P_{33} \\ P_{34} \\ P_{44} \end{bmatrix} = \frac{\nu}{4\delta k(\delta^2+\omega^2)} \begin{bmatrix} 2\delta^2+\omega^2 \\ \delta\omega \\ \omega^2 \\ 0 \\ \hdashline (2\delta^2+\omega^2)k \\ \delta\omega k \\ \omega^2 k \end{bmatrix} . \tag{7.79}$$

Thus, the result (7.76) is completely confirmed. In this example, the implicit solution seems to be more adequate than the explicit solution. However, this is not a general statement.

APPENDIX A

Controllability and Observability

The concepts of controllability and observability are two of the most essential concepts in modern dynamical system theory. They are important to the design of control systems and give an insight into the physical problem. Roughly speaking, controllability characterizes the influence of input forces to the dynamical behavior of the system while observability characterizes the information on the state by output measurement. If a dynamical system is completely controllable, all the eigenmodes of the system can be excited from the input; if a dynamical system is completely observable, all the eigenmodes of the system can be observed at the output. Assume a linear time-invariant dynamical system represented by a state equation

$$\dot{\mathbf{x}}(t) = \mathbf{A}\mathbf{x}(t) + \mathbf{B}\mathbf{u}(t) \tag{A.1}$$

and by an output equation

$$\mathbf{z}(t) = \mathbf{C}\mathbf{x}(t) \tag{A.2}$$

where $\mathbf{x}$, $\mathbf{u}$, and $\mathbf{z}$ denote state, input, and output vectors of dimensions n, r, s, respectively. The matrices $\mathbf{A}$, $\mathbf{B}$, $\mathbf{C}$ are of order compatible with the vectors $\mathbf{x}$, $\mathbf{u}$, and $\mathbf{z}$. Then, it holds the following definitions.

Definition A.1: Complete controllability

The dynamical system (A.1) is said to be completely (state) controllable if for any initial state $\mathbf{x}(t_0) = \mathbf{x}_0$ and any state $\mathbf{x}_1$ there exists a finite time $t_1 > t_0$ and an input $\mathbf{u}(t)$ defined on $[t_0, t_1]$ that will transfer the state $\mathbf{x}_0$ to the state $\mathbf{x}_1$ at time t_1.

This definition requires only that the input $\mathbf{u}$ is capable of moving any state in the state space to any other state in a finite time; what trajectory the state should take is not specified. Furthermore, there is no constraint imposed on the input; its magnitude can be as large as desired.

Definition A.2: Complete observability

The dynamical system (A.1, A.2) is said to be completely observable if for any state $\mathbf{x}(t_0) = \mathbf{x}_0$ there exists a finite time $t_1 > t_0$ such that the knowledge of the input $\mathbf{u}(t)$ and the output $\mathbf{z}(t)$ over the time interval $[t_0, t_1]$ suffices to determine the state $\mathbf{x}_0$.

General criteria for controllability and observability of linear time-invariant systems are due to Kalman (1959/60).

Theorem A.1: Kalman criterion for controllability

The dynamical system (A.1) is completely controllable if and only if

(A.3) $$\operatorname{rank}\left[\mathbf{B}\,|\,\mathbf{AB}\,|\,\ldots\,|\,\mathbf{A}^{n-1}\mathbf{B}\right] = n\,.$$

Theorem A.2: Kalman criterion for observability
The dynamical system (A.1, A.2) is completely observable if and only if

(A.4) $$\operatorname{rank}\left[\mathbf{C}^T\,|\,\mathbf{A}^T\mathbf{C}^T\,|\,\ldots\,|\,\mathbf{A}^{T^{n-1}}\mathbf{C}^T\right] = n\,.$$

Uncontrollable or unobservable states $\tilde{\mathbf{x}}$ are characterized by

$$\tilde{\mathbf{x}}^T\left[\mathbf{B}\,|\,\mathbf{AB}\,|\,\ldots\,|\,\mathbf{A}^{n-1}\mathbf{B}\right] = 0$$

or

$$\tilde{\mathbf{x}}^T\left[\mathbf{C}^T\,|\,\mathbf{A}^T\mathbf{C}^T\,|\,\ldots\,|\,\mathbf{A}^{T^{n-1}}\mathbf{C}^T\right] = 0\,,$$

respectively.
Further convenient controllability and observability criteria are due to Hautus (1969) who gives a nice relationship of these concepts to the eigenmodes of the system matrix $\mathbf{A}$.

Theorem A.3: Hautus criterion for controllability
The dynamical system (A.1) is completely controllable if and only if each right eigenvector of the transposed system matrix $\mathbf{A}^T$ is not orthogonal to all columns of $\mathbf{B}$, or equivalently

(A.5) $$\mathbf{A}^T\mathbf{x}_i = \lambda_i\mathbf{x}_i \text{ implies } \mathbf{B}^T\mathbf{x}_i \neq 0$$

for all right eigenvectors $\mathbf{x}_i$ of $\mathbf{A}^T$.
On the contrary, $\mathbf{A}^T\mathbf{x}_i = \lambda_i \mathbf{x}_i$ with $\mathbf{B}^T\mathbf{x}_i = \mathbf{0}$ means that the i-th eigenmode of the homogeneous system $\dot{\mathbf{x}}(t) = \mathbf{A}\mathbf{x}(t)$ is not excited by any control $\mathbf{B}\mathbf{u}(t)$. Therefore, this eigenmode (and also the system) is not controllable.

Theorem A.4: Hautus criterion for observability
The dynamical system (A.1, A.2) is completely observable if and only if each right eigenvector of the system matrix $\mathbf{A}$ is not orthogonal to the rows of $\mathbf{C}$:

$$\mathbf{A}\mathbf{x}_i = \lambda_i \mathbf{x}_i \quad \text{implies} \quad \mathbf{C}\mathbf{x}_i \neq \mathbf{0} \tag{A.6}$$

for all right eigenvectors $\mathbf{x}_i$.
Does not hold relation (A.6) for one $\mathbf{x}_i$ this eigenmode is not observable by the output $\mathbf{z}(t)$. Therefore the system is not completely observable.

APPENDIX B

Lyapunov Matrix Equation

In this appendix the Lyapunov matrix equation is considered

$$\mathbf{A}^T\mathbf{R} + \mathbf{R}\mathbf{A} = -\mathbf{S} \tag{B.1}$$

where $\mathbf{A}$, $\mathbf{S}$ are given $n \times n$ matrices and $\mathbf{R}$ is an unknown $n \times n$ matrix. This equation arises in various fields of linear system theory: stability analysis, optimization theory, calculation of covariance matrices, investigation of sensitivity of control systems. Therefore, it is necessary to look for existence and uniqueness of the solutions of (B.1) as well as for analytical and numerical methods for solving (B.1).

Theorem B.1: Existence
For given (real) matrices $\mathbf{A}$ and $\mathbf{S}$ a (real) solution $\mathbf{R}$ of (B.1) exists if and only if there is at least one (real) similarity transformation matrix $\mathbf{T}$ such that

$$\mathbf{T}^{-1}\begin{bmatrix} \mathbf{A} & \mathbf{0} \\ -\mathbf{S} & -\mathbf{A}^T \end{bmatrix}\mathbf{T} = \begin{bmatrix} \mathbf{A} & \mathbf{0} \\ \mathbf{0} & -\mathbf{A}^T \end{bmatrix}. \tag{B.2}$$

This theorem is due to Roth (1952) and is plausible by the relation

$$\begin{bmatrix} E & 0 \\ R & E \end{bmatrix}^{-1} \begin{bmatrix} A & 0 \\ -S & -A^T \end{bmatrix} \begin{bmatrix} E & 0 \\ R & E \end{bmatrix} = \begin{bmatrix} A & 0 \\ -(A^T R + RA + S) & -A^T \end{bmatrix} = \begin{bmatrix} A & 0 \\ 0 & -A^T \end{bmatrix} \tag{B.3}$$

if $\mathbf{R}$ is a solution of (B.1).

Theorem B.2: Uniqueness

If λ_i, $i = 1(1)n$, are the eigenvalues of the matrix $\mathbf{A}$, then (B.1) has a unique solution $\mathbf{R}$ if and only if

$$\lambda_i + \lambda_j \neq 0 \tag{B.4}$$

for all $i, j = 1(1)n$. This requires that $\mathbf{A}$ has no zero eigenvalue and no eigenvalues which are opposite, or equivalently, that $\mathbf{A}$ and $-\mathbf{A}$ do not have common eigenvalues. The condition (B.4) is also equivalent to

$$H_n = \det \mathbf{H} \neq 0 \tag{B.5}$$

where $\mathbf{H}$ is the Hurwitz matrix (4.14, 4.15).

This is a very well-known theorem and its proof is given e.g. in the book of Lancaster (1969). There is also the next theorem proven.

Theorem B.3: Integral solution

If all eigenvalues of the matrix $\mathbf{A}$ have negative real parts, then the unique solution (B.1) is given by

(B.6) $$\mathbf{R} = \int_0^\infty e^{\mathbf{A}^T\tau}\,\mathbf{S}\,e^{\mathbf{A}\tau}\,d\tau .$$

This result is easily verified. The infinite integral converges and

$$\mathbf{A}^T\mathbf{R} + \mathbf{R}\mathbf{A} = \int_0^\infty \left[\mathbf{A}^T e^{\mathbf{A}^T\tau}\mathbf{S}e^{\mathbf{A}\tau} + e^{\mathbf{A}^T\tau}\mathbf{S}e^{\mathbf{A}\tau}\mathbf{A}\right] d\tau =$$

$$= \int_0^\infty \frac{d}{d\tau}\left[e^{\mathbf{A}^T\tau}\mathbf{S}e^{\mathbf{A}\tau}\right] d\tau = e^{\mathbf{A}^T\tau}\mathbf{S}e^{\mathbf{A}\tau}\Big|_0^\infty = -\mathbf{S} .$$

In the opposite to theorem B.3 where asymptotic stability of the system matrix $\mathbf{A}$ is assumed, the following analytical solution yields always if the solution is unique as shown by Müller (1969).

Theorem B.4: Leverrier-Faddeyev algorithm solution
Let the characteristic coefficients $\mathbf{A}$ of the system matrix given by the Leverrier-Faddeyev algorithm,

(B.7)
$$a_k = -\frac{1}{k}\,\mathrm{tr}\,\mathbf{A}\mathbf{A}_{k-1} , \quad a_0 = 1 , \quad \mathbf{A}_0 = \mathbf{E} ,$$
$$\mathbf{A}_k = \mathbf{A}\mathbf{A}_{k-1} + a_k\mathbf{E} , \quad k = 1(1)n ,$$

and let $\mathbf{q}$ be a solution of

(B.8) $$\mathbf{H}\mathbf{q} = \frac{1}{2}\,\mathbf{e}_1$$

where $\mathbf{H}$ is the nonsingular Hurwitz matrix (4.14) $(H_n = \det \mathbf{H} \neq 0)$ and $\mathbf{e}_1$ the first unit vector, then

$$\mathbf{R} = \sum_{j=0}^{n-1} q_{j+1} \sum_{i=0}^{2j} (-1)^i \mathbf{A}_i^T \mathbf{S} \mathbf{A}_{2j-i} \tag{B.9}$$

is the unique solution of (B.1) (remark: $\mathbf{A}_k = \mathbf{0}$ for $k \geq n$).

The solution (B.6) is generally only of theoretical interest while (B.9) is very useful for the solution (B.1) in the case of low order systems. But when the system order increases one has to look for numerically stable solution algorithms for the Lyapunov matrix equation (B.1). Here we will give three successfully tested computer algorithms.

Theorem B.5: Direct solution method

Assuming $\mathbf{S}$ is a symmetric matrix, $\mathbf{S} = \mathbf{S}^T$, the solution matrix $\mathbf{R}$ of (B.1) is symmetric, too, $\mathbf{R} = \mathbf{R}^T$, and consists of $n(n+1)/2$ unknown variables. By organizing $\mathbf{R} = \mathbf{R}^T$ and $\mathbf{S} = \mathbf{S}^T$ as vectors

$$\mathbf{r} = \left[R_{11}\, R_{12} \cdots R_{1n}\, R_{22} \cdots R_{2n}\, R_{33} \cdots R_{nn}\right]^T,$$

$$\mathbf{s} = \left[S_{11}\, S_{12} \cdots S_{1n}\, S_{22} \cdots S_{2n}\, S_{33} \cdots S_{nn}\right]^T, \tag{B.10}$$

the system (B.1) is rewritten as common linear equations

$$\mathcal{A}\,\mathbf{r} = \mathbf{s}, \tag{B.11}$$

and can be solved by general methods, like Gaussian elimination. The matrix $\mathscr{A}$ can be formed from $\mathbf{A}$ by use of logical operations, Chen and Shieh (1968).

The programming of this algorithm is easily done. The main disadvantages are that the memory requirement is $[n(n+1)/2]^2$ and the number of multiplicative operations for large n is of the order $n^6/24$. But for small systems (say: $n < 8$) it is the best algorithm.

Theorem B.6: Iterative, infinite series solution method
Assuming $\mathbf{A}$ is an asymptotically stable matrix $(\operatorname{Re} \lambda_i < 0)$ then the solution of (B.1) is given by

(B.12)
$$\mathbf{R} = \sum_{k=1}^{\infty} \mathbf{V}^{T^{k-1}} \bar{\mathbf{S}} \mathbf{V}^{k-1}$$

where

(B.13)
$$\bar{\mathbf{S}} = 2a(a\mathbf{E} - \mathbf{A}^T)^{-1} \mathbf{S} (a\mathbf{E} - \mathbf{A})^{-1}$$

and

(B.14)
$$\mathbf{V} = (a\mathbf{E} + \mathbf{A})(a\mathbf{E} - \mathbf{A})^{-1}.$$

The positive real number $a > 0$ is chosen such that $(a\mathbf{E} - \mathbf{A})$ is a well conditioned nonsingular matrix and that the infinite series solution (B.12) has a good convergence rate. A good convergence is yield by

$$a \approx -\frac{1}{n}\,\mathrm{tr}\,\mathbf{A} = \frac{a_1}{n}$$

or (B.15)

$$a \approx \sqrt[n]{(-1)^n \det \mathbf{A}} = \sqrt[n]{a_n} \; .$$

Accelerated convergence of (B.12) is obtained by a suggestion of Smith (1968):

$$\mathbf{R}_{k+1} = \mathbf{R}_k + \mathbf{V}^{T^{2^k}} \mathbf{R}_k \mathbf{V}^{2^k} \; , \quad \mathbf{R}_0 = \bar{\mathbf{S}} \; ,$$

$$\lim_{k \to \infty} \mathbf{R}_k = \mathbf{R} \; . \tag{B.16}$$

Since $\mathbf{V}^{2^k}$ is obtained by squaring the matrix $\mathbf{V}^{2^{k-1}}$ the iterative procedure (B.16) is convenient for computation. For large order systems (say: $n > 8$) and asymptotically stable matrices the iterative infinite series solution (B.16) can be recommended.

Theorem B.7: Solution via Hessenberg form

Let $\mathbf{A}_H$ be a lower Hessenberg representation of $\mathbf{A}$, which can be computed efficiently by the use of a numerically stable Hessenberg process, Zurmühl (1964): $\mathbf{T}^{-1}\mathbf{A}\mathbf{T} = \mathbf{A}_H$.

After that (B.1) reads as

$$\mathbf{A}^T \tilde{\mathbf{R}} + \tilde{\mathbf{R}} \mathbf{A}_H = -\tilde{\mathbf{S}} \tag{B.17}$$

where

$$\text{(B.18)}\quad \tilde{\mathbf{R}} = \mathbf{R}\mathbf{T}\,, \quad \tilde{\mathbf{S}} = \mathbf{S}\mathbf{T}\,, \quad \mathbf{A}_H = \begin{bmatrix} h_{11} & h_{12} & & 0 \\ h_{21} & h_{22} & h_{23} & \\ \vdots & & \ddots & \\ h_{n-1,1} & \cdots & \cdots & h_{n-1,n} \\ h_{n1} & \cdots & \cdots & h_{nn} \end{bmatrix} .$$

Then the columns $\tilde{\mathbf{R}}_i$ of $\tilde{\mathbf{R}}$ are obtained by the algorithm

$$\text{(B.19)}\qquad \tilde{\mathbf{R}}_n = \mathbf{G}_H^{-1}\,\mathbf{d}_H\,,$$

$$\text{(B.20)}\qquad \tilde{\mathbf{R}}_{i-1} = -\frac{1}{h_{i-1,i}}\left(\mathbf{A}^T\tilde{\mathbf{R}}_i + \sum_{j=1}^{n} h_{ji}\,\tilde{\mathbf{R}}_j + \tilde{\mathbf{S}}_i\right), \quad i = n(1)1\,,$$

where $\tilde{\mathbf{S}}_i$ are the columns of $\tilde{\mathbf{S}}$, and $\mathbf{d}_H$ and $\mathbf{G}_H$ can be calculated using the recursive algorithm (B.20) as follows: substitute $\tilde{\mathbf{R}}_n = \mathbf{0}$ in (B.20), then for $i = 1$ $\tilde{\mathbf{R}}_0 = -\mathbf{d}_H$ is obtained, whereas for $\tilde{\mathbf{S}} = \mathbf{0}$ and $\tilde{\mathbf{R}}_n = \mathbf{e}_k$ the algorithm (B.20) yields for $i = 1$ the k-th column of $\mathbf{G}_H$, $k = 1(1)n$.

This result is due to Kreisselmeier (1972), but a similar result is also given by Meyer-Spasche (1972). The procedure is recommended if the application of theorems B.5 or B.6 is not possible or if one is interested in the solution of the Lyapunov matrix equation as well as in the eigenvalue-eigenvector problem. In both problems the transformation to a Hessenberg representation is needed. The numerical process of theorem B.7 can be made very stable, especially roundoff errors in the algorithm (B.19, B.20) can be reduced by a simple repeated implementation of the algorithm itself (see Kreisselmeier (1972)).

Further comparison of numerical methods for solving the Lyapunov matrix equation can be found in the recent papers of Hagander (1972) and of Pace and Barnett (1972), (where, however, the new algorithm of theorem B.7 is not taken into account).

REFERENCES

L. Cesari (1963): "Asymptotic Behavior and Stability Problems in Ordinary Differential Equations". Ergebnisse der Mathematik und ihrer Grenzgebiete, Neue Folge, Bd. 16, 2. Auflage, Springer-Verlag, Berlin-Göttingen-Heidelberg.

C.F. Chen and L.S. Shieh (1968): "A Note on Expanding $PA+A^TP=-Q$." IEEE Transactions on Automatic Control 13, pp. 122-123.

S.H. Crandall and W.D. Mark (1963): "Random Vibrations in Mechanical Systems", Academic Press, New York.

R.F. Drenick (1967): "Die Optimierung linearer Regelsysteme", Oldenbourg, München-Wien.

L. Fabian (1973): "Zufallschwingungen und ihre Behandlung", Springer-Verlag, Berlin-Heidelberg-New York.

P. Hagander (1972): "Numerical Solution of $A^TS+SA+Q=0$." Information Sciences 4, pp. 35-50.

M.L.J. Hautus (1969): "Controllability and Observability Conditions of Linear Autonomous Systems", Indagationes Mathematicae 31, pp. 443-448.

K. Itô (1944): "Stochastic Integral", Proc. Imp. Acad. Tokyo 20, pp. 519-524.

H.M. James, N.B. Nichols and R.S. Phillips (1947): "Theory of Servomechanisms", McGraw-Hill, New York.

A.H. Jazwinski (1970): "Stochastic Processes and Filtering Theory", Academic Press, New York-London.

R.E. Kalman (1960a): "Contributions to the Theory of Optimal Control", Proc. Mexico City Conf. Ordinary Differential Equations, 1959, Bole.Soc.Math. Mexicana 5, pp. 102-119.

R.E. Kalman (1960b): "A New Approach to Linear Filtering and Prediction Problems", J.Basic Eng. ASME, 82, pp. 35-45.

K. Klotter (1951): "Technische Schwingungslehre", Bd. 1, Einläufige Schwinger, 2. Auflage, Springer-Verlag, Berlin-Heidelberg-Göttingen.

K. Klotter (1960): "Technische Schwingungslehre", Bd. 2, Mehrläufige Schwinger, 2. Auflage, Springer-Verlag, Berlin-Heidelberg-Göttingen.

V. Krebs (1973): "Das Gleichgewichtstheorem - eine grundsätzliche Aussage über das Verhalten von Regelkreisen", Teil 1 und 2. Regelungstechnik und Prozess-Datenverarbeitung 21, pp. 25-27 and 56-59.

G. Kreisselmeier (1972): "A Solution of the Bilinear Matrix Equation $AY+YB=-Q$", SIAM J. Appl. Math. 23, pp. 334-338.

H. Kwakernaak and R. Sivan (1972): "Linear Optimal Control Systems", Wiley-Interscience, New York.

P. Lancaster (1969): "Theory of Matrices", Academic Press, New York.

S.H. Lehnigk (1966): "Stability Theorems for Linear Motions with an Introduction to Lyapunov's Direct Method", Prentice Hall, Englewood Cliffs, N. J.

P.W. Likins (1971): "Lectures on Dynamics of Flexible Spacecraft", International Centre for Mechanical Sciences (CISM), Courses and Lectures N° 103, Udine.

H. Lippmann (1968): "Schwingungslehre", Hochschultaschenbücher Nr. 189/189a, Bibliograph. Institut Mannheim.

K. Magnus (1969): "Schwingungen", 2. Auflage, Teubner, Stuttgart.

K. Magnus, J. Lückel, P.C. Müller and W. Schiehlen (1971): "Moderne Schwingungstechnik", Institut B für Mechanik, München.

L. Meirovitch (1967): "Analytical Methods in Vibrations", MacMillan, New York.

R. Meyer-Spasche (1972): "A Constructive Method of Solving the Ljapunov Equation for Complex Matrices" Numer. Math. 19, pp. 433-438.

P. Müller (1969): "Die Berechnung von Liapunov-Funktionen und von quadratischen Regelflächen für lineare stetige zeitinvariante Mehrgrössensysteme", Regelungstechnik 17, pp. 341-345.

P.C. Müller (1970): "Special Problems of Gyrodynamics" International Centre for Mechanical Sciences (CISM) Courses and Lectures N° 63, Udine.

P.C. Müller (1974): "Matrizenverfahren in der Stabilitätstheorie linearer dynamischer Systeme", Habilitationsschrift, Technische Universität München.

I.S. Pace and S. Barnett (1972): "Comparison of Numerical Methods for Solving Liapunov Matrix Equations", Int. J. Control 15, pp. 907-915.

A. Papoulis (1965): "Probability, Random Variables and Stochastic Processes", McGraw-Hill, New York.

H. Parkus (1969): "Random Processes in Mechanical Sciences", International Centre for Mechanical Sciences (CISM), Courses and Lectures N° 9, Udine.

W.E. Roth (1952): "The Equations $AX-YB=C$ and $AX-XB=C$ in Matrices", Proc. Amer. Math. Soc. 3, pp. 392-396.

W. Schiehlen (1972): "Zustandgleichungen elastischer Satelliten", Z. angew. Math. Phys. 23, pp. 575-586.

W.O. Schiehlen (1973): "A Fine Pointing System for the Large Space Telescope", NASA TN D-7500, Washington D.C.

W. Schiehlen (1974): "Zur Untersuchung von Zufallschwingungen", Z. angew. Math. Mech. 54, pp. T64-T65.

R.A. Smith (1968): "Matrix Equation $XA+BX=C$", SIAM J. Appl. Math. 16, pp. 198-201.

N. Wiener (1949): "The Extrapolation, Interpolation and Smoothing of Stationary Time Series", Wiley, New York.

J.L. Willems (1970): "Stability Theory of Dynamical Systems", Nelson, London.

O.C. Zienkiewicz (1971): "The Finite Element Method in Engineering Science", 2nd edition, McGraw-Hill, London.

R. Zurmühl (1964): "Matrizen und ihre technische Anwendungen", 4. Auflage, Springer Verlag, Berlin-Göttingen-Heidelberg.

CONTENTS

Printed in the United States
By Bookmasters